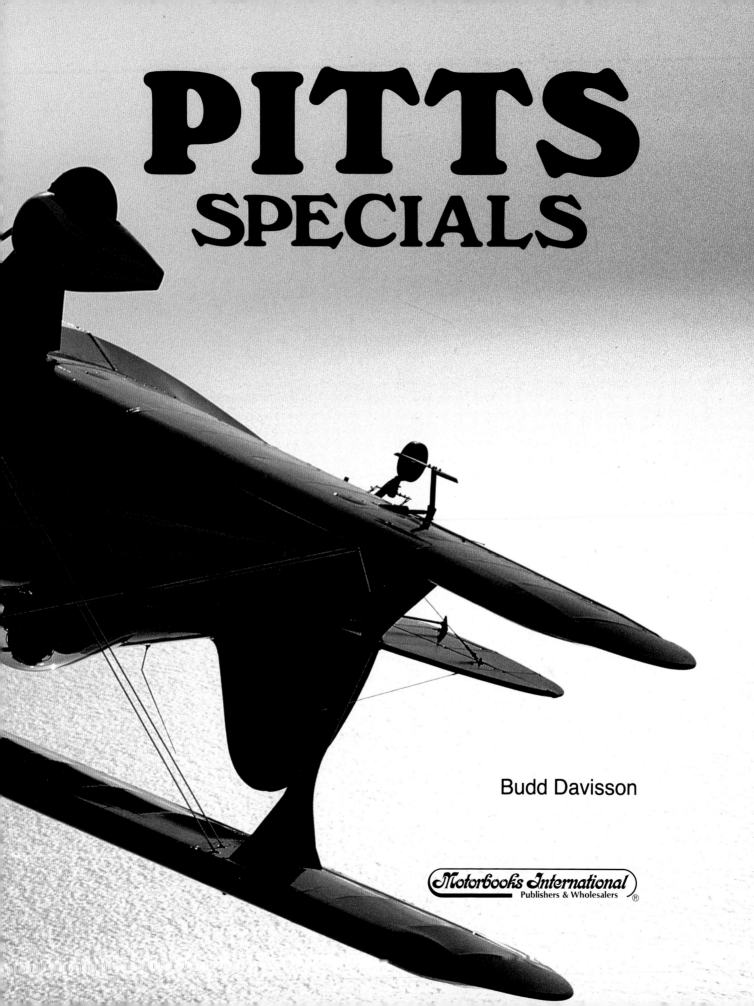

PITTS
SPECIALS

Budd Davisson

Motorbooks International
Publishers & Wholesalers

To Curtis, for adding a level of artistic excitement to our lives that makes us proud to be known as Pitts pilots.

To Willie Mae Pitts, for putting up with, and supporting, Curtis, which made it all possible. We owe you a big one, Ma.

First published in 1991 by Motorbooks International Publishers & Wholesalers, P O Box 2, 729 Prospect Avenue, Osceola, WI 54020 USA

Motorbooks International books are also available at discounts in bulk quantity for industrial or sales-promotional use. For details write to Special Sales Manager at the Publisher's address

Library of Congress Cataloging-in-Publication Data
Davisson, Budd.
    Pitts specials / Budd Davisson.
      p.   cm.
    Includes index.
    ISBN 0-87938-536-7
    1. Pitts aircraft.   I. Title.
TL686.P54D38   1991
797.5'4—dc20   91-12404

Printed and bound in Hong Kong

On the front cover: *Bill Thomas' S–1S Pitts was a frequent competitor in the world aerobatic contests of the late 1960s and early 1970s.*

On the back cover: *A sample of the many Pitts aircraft from the book: an S–2A, an S–1T, the replica of* Sampson *(photo by S. Wolf) and an S–2A from Manx Kelly's Carling aerobatic team.*

On the frontispiece: *Pitts pilots don't have to worry about gravity, as this formation of pilots from Aero Sport in Saint Augustine, Florida, demonstrates.*

On the title page: *Pitts Specials are the ultimate freedom machines.*

Contents

Acknowledgments

I need to thank a lot of folks for their help in getting this book finished and, if I forget someone, they know how absentminded I am, so hopefully they won't be too upset.

Jack and Golda Cox have to be singled out for having the patience to put up with my never-ending phone calls for information. They have all the answers, so they get all the calls. The rest of the folks up at the EAA, from Tom Poberezny on down were super-helpful.

Verdean Heiner and Herb Anderson at Christen Industries (now Aviat) filled in a lot of blanks and loaned me some important graphic material.

Carol Buggeln, my faithful organizer, typist and right hand, is to be commended for producing a manuscript filled with words she had never heard before. Now she wants a Pitts of her own.

The photos would never have happened if it hadn't been for the guys at Aero Sport in Saint Augustine, Florida. Carl "Splash" Pascarell, Big Jim Moser, "Glider" John Homroc, Eliot Cross and Craig Fordem all flew their biplane buns off for me to get Pitts pictures. It takes very special pilots to do the things we were doing and come away with consistently high quality shots. The proper recipe for good air-to-air photography is, and always will be, 70 percent pilots, 20 percent weather, 5 percent luck and 5 percent photographer. If it isn't there to shoot, the best photographer in the world isn't going to get the shot.

And then there is Curtis Pitts. He sat with me on the phone for hours while we tried to get it all down right. And I probably still screwed up. If so, I hope they were minor mistakes.

Introduction

It would be easy to begin this tale by painting a picture of a dusty, red clay road winding its way through sparse slash pines. The southern Georgia sun has converted the clay to powder which rises in a small silent rooster tail behind the bicycle of young Curtis Pitts, as he pedals toward Souther Air Field. It would be easy to use brush strokes composed of words to conjure up a mental image of 1927 Americus, Georgia, to set the scene for young Pitts and his entry into the hallowed world of the aviator.

Although an accurate description, that's not the image the reader should have in mind. Although rural Georgia is an important part of the story, the central essence, the underlying emotion, should spring from a far different source. The image perceived should be one of sitting, feet spread wide apart in a tiny, Spartan steel tube and fabric cocoon that is totally filled with noise. And born of motion. The image should be one of looking out at a whirling world through a tiny eyelid of plexiglass meant to deflect the on-rushing wind, as it is cleaved aside by the blunt nose and collection of wings, wires and struts.

The image in the reader's mind should reside in a wildly moving machine that encases him like a suit of chain mail; but, like a warrior in combat, he is totally unaware of its existence. In fact, it is not a matter of man and machine, since, with this particular machine, the Pitts Special, the biomechanical line of demarcation between the two becomes blurred. It's hard to tell where one begins and the other ends. The reader should experience a feeling of effortless connection, where neurons flow directly from the cerebral cortex out into the slipstream where they are converted to flight in a heartbeat.

This is not a story about an airplane. It is not a story about a man. It is not a story about aerobatics, or custom-crafted flying machines. It is the complex tale of the intangible, emotional and thoroughly illogical activity known as flight, and what a single man and an overdeveloped, overengineered and overloved little machine have contributed to it.

It is a story of people. Of nuts and bolts. Of a brutally artistic sort of fun. But, all the while, keep the proper image in mind. Think of being able to tie the horizon into knots with a mere flicker of your mind, and an immeasurable twitch of your wrist. Think of any movement you've ever observed in nature, be it a darting fish, or more appropriately, a frolicking, smiling otter, and know you can emmulate it. Know that the bi-winged cocoon removes all earthly limitations. Know that you are as close to total, three-dimensional freedom as you will ever get.

In bumpersticker language, the central theme of this saga might be, "Freedom, thy name is Pitts Special."

Bill Thomas' S–1S Roundwing was a familiar sight during the World contests of the late 1960s and early 1970s. It typified the Pitts that beat the best the world had to offer.

Curtis and Willie Mae Pitts, better known as Ma and Pa to aerobatic pilots everywhere. This photo was taken at the Pitts Party in 1990.

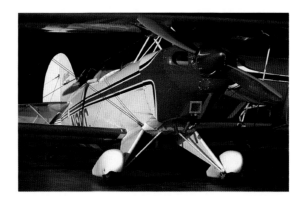

A Man Named Curtis

In 1927, rural towns the size of Americus, Georgia, should have had no airport. Not a proper one, anyway. Usually, a tiny piece of pasture land became a favorite alighting place for the occasional barnstormer, but that was it. Americus, however, was luckier than most. It was one of only a few towns in the United States to have received a training field courtesy of the US government and at the insistence of Kaiser Wilhelm. As Curtis Pitts pedaled his way out to Souther Field, only a decade had passed since the broad expanse of grass had been lined with Curtiss Jennys, the air cluttered with the clatter of slow-moving OX-5s, as America strived to build its own air force.

In the decade after the war, time and the loss of government money had taken its toll on Souther. Still, the basic facilities were there. The tumbled-down hangars, the runways and, most important, the interest in aviation which had been planted by the intense activity in 1917-18 were still alive. For the boy Curtis Pitts, the airfield was a magnet.

Curtis arrived in Americus as an orphan from the tiny town of Stillmore on the other side of the state. Even today, as time creeps toward a new century, Stillmore, Georgia, boasts a population of only 527 hard-working folks.

Pitts like this S–2A always look ready to go play, even while sitting in a hangar.

Industry has changed little from the times when Drew Lee Pitts and his wife, Amanda, had a small red-dirt farm a few miles southeast of town where they lived while Drew worked in the sawmills. There was nothing to hint of aeronautical beginnings.

By the time Curtis was ten years old, he had lost both of his parents to the hard life that was part of rural Georgia. With no relatives in the immediate area, he moved to Americus to be brought up by his uncle, Lester Dickerson, a railroad bridge foreman.

Curtis was a typical teenager, regardless of the time period: If it wasn't cars, girls, girls in cars, or sports, he couldn't be concerned with it. However, he had one other interest not common among Georgian teenagers—airplanes. To use his own words, "Airplanes got a hold of me real early and just never seemed to let go. . . ."

Love affairs all have beginnings and his began in earnest one afternoon, when he was thirteen years old. He and his friend, Clarence Niblack, decided it was time to see where those flying machines they had seen overhead were roosting, so they sneaked out to the airport, five miles down that dusty road. It wasn't exactly forbidden territory, but they also knew their parents weren't going to be crazy about them going so far by themselves. Still, it was a calculated risk and they had to get out to that aerial oasis to see what they could see.

They hadn't been at the airport long when Niblack's father and Curtis' uncle showed up. They were a bit ruffled, and the boys were both convinced they were going to get a licking. In those days a Georgian licking was something to be feared. The elder Niblack and the uncle huddled together just out of earshot of the two boys, both of whom were expecting the worst. Then, with no explanation, the adults turned and approached pilot Frank Bentley, paying him to take the boys for a ride in his WACO 9.

Sixty years after the fact, Curtis is still not sure why Niblack's father changed his mind and went for his wallet instead of his belt. "I was positive we were in for it. I wonder if Clarence's father had known what he was getting me in for in the long run, if he would have thought those few bucks were a good investment or not," says Pitts.

As Curtis describes it, the first ride was memorable, if nothing else, because as they were strapping in together in the front of the WACO, a third boy came clambering in and they couldn't get him out. So, Curtis Pitts took his first airplane ride as part of a trio shoehorned into the front seat of a WACO 9.

Curtis admits to not setting the academic world on fire in high school. Academics seemed too slow and dead for him. He spent as much time as possible hanging around the airport watching John Wyche and his crew cutting and gluing, stitching and doping on the machines he saw as absolute works of art. Wyche ran a part-time aircraft repair service and could lay claim to having sold a tall guy named Lindbergh his first Jenny right there at that airport, a fact he let no one forget after 1927.

"The airport didn't have a whole lot of airplanes based there, but lots of barnstormers came by for repair work and spent several weeks at a time. One of those, Frank Bentley, had the nearly new, straight-axle WACO 9 I had gotten the ride in and the machine was pure magic, to me," remembers Pitts.

In school he tried to keep his mind on reading and writing, but his mind was somewhere else, tied up in rib-stitching and welding. In fact, he learned to weld while still in high school, a skill that would, in later years, prove valuable to him.

He remembers his uncle as being an easygoing, affable individual who liked nothing better than taking his nephew hunting or fishing. He also tried to understand the young Pitts' fascination with flying machines. Pitts' Aunt Minona, however, didn't agree with that view. She made no bones about not liking Curtis hanging around the airport, probably for the same reason she wouldn't let him play football—she was afraid he'd get hurt.

Curtis laughs a lot when he talks about his childhood. In fact, he laughs a lot in general. His normal conversational tone rumbles along in a deep drawl, constantly on the edge of an even deeper chuckle. The listener knows there is something serious about to be learned when his tone of voice steps a millimeter away from the chuckle and slides into a slightly somber tone for a second or two. Then it's back to the laughter.

Curtis Pitts has always been able to see the fun side of life, a quality that developed very early, probably during those hours he spent as the stereotypical airport kid. It was during that same period he became totally enamored of the nuts and bolts which hold life together. When he was not even halfway through high school, he began to see the structure of aircraft as something that was not only understandable, but something that he felt even his young hands could master. Between burying his nose in *Modern Mechanics*, soaking up all the information contained in their yearly twenty-five-cent flying manuals, and the hands-on education he was getting out at the airport, he decided it was time to try his hand at creating a flying machine.

Money for his aerial daydreams came from a wide series of sources, all of them part-time jobs. He delivered papers all over town and served as a projectionist in the local theater. As a carpenter's helper he was making twenty-two cents an hour, which was a princely wage for a young high schooler in depression-era Georgia.

The first Pitts flying machine borrowed heavily on the design of Heath and Pietenpol Parasols, and used a wheezing Model T Ford engine. On a good day that old Ford cranked out 20 horsepower. On a bad day it just wheezed. The propeller was off of a 28 hp Lawrence and did its best to blow enough breeze back to keep the airplane moving.

The airplane was constructed in the garage of the house next door to Curtis, which was also owned by his uncle. Adjacent to the houses was a large square, smooth grassy area that quickly became Curtis Pitts' private aerodrome.

"As soon as I got that thing running," remembers Pitts, "I'd go out in that field and run around in circles. I practically taxied the wings off of it."

Eventually, during one of Pitts' exuberant laps of his private world, a gust of wind got under the wing and literally rolled the airplane into a ball. This could have been a blessing in disguise, considering Pitts hadn't had a single minute of flight instruction yet.

After he had moved out of town, he received a letter from his old friend and first-time flying buddy, Clarence Niblack, saying he had gotten an offer of $6 for the remains. "I wrote him back and told him to send me the $6, which he did. That may have been one of the best airplane deals I've made to date," says Curtis.

The class of Americus, Georgia, 1934 graduated into a less than hospitable economic world and jobs were hard to find everywhere. The local economy couldn't absorb anywhere near the number of graduates, so young Curtis packed his bags and headed off to the closest major town, Jacksonville, Florida, where his uncle had gotten him a job with the railroad.

Even though he was a young man, his hands were much more experienced than his age showed. The time he had spent learning to build airplanes had taught him how to build anything, a talent Seaboard Rail Lines needed. So they hired him on as a carpenter to work on their road gangs.

Paid twenty-two-and-a-half cents an hour, Curtis was part of a carpentry crew that rode up and down the various legs of the railroad's routes, stopping to build and repair wooden structures needing attention. Riding a special train which included a dormitory car, a cook car and several for hauling building materials, he saw a lot of Georgia and Florida over the next few years. The most important construction job they undertook, in terms of its impact on his future, was repairing the depot and packing houses in Ocala, Florida.

Living in a dormitory car on a siding, the combination of free time and free room and board gave the eighteen-year old some important ingredients for his future life: time and money.

Curtis recalls: "I was making darn good money for the time and on that job in Ocala I really was making more than I needed. That's when I decided it was time to take flying lessons."

On the northwest side of Ocala was Taylor Field. A municipal airport, it was typical of WPA airports in that it consisted of several hangars, a few airplanes and about eighty acres of grass. Most important, it had a small flying school operated by Doc Chiddix with an E-2 Cub. Although Chiddix was only a private pilot who couldn't technically solo Curtis, for $3 a half hour he and Curtis would go out and explore the educational experience together. Since he could ride the railroad for free, Curtis began to spend every weekend at Ocala taking flying lessons.

"Doc was a reasonably good instructor, but it was the E-2 that really taught me everything," Curtis remembers of his training days.

After Chiddix had gotten him just so far along, one of the local instructors, Bob Wimple, jumped in with him and made sure he knew what he was doing before turning him loose. At the time Curtis Pitts soloed, he had four hours and twenty minutes in his slim little logbook. That was July of 1934.

In the summer of 1938 Pitts was working in Baldwin, Florida, when a friend invited him over for dinner. As Pitts entered the house, he met the real reason for the invitation: Willie Mae Lord. It was a setup by his friend and it worked. Although Curtis says he didn't think much of it at the time, he will admit to immediately taking off for the swimming hole up the road with Willie Mae. The friend's matchmaking worked; Willie Mae and Curtis were married in October of that same year. From that day forward, she was an integral part of the Pitts formula and was one of Curtis' staunchest supporters.

Increasingly, Curtis began to lead a double life: one as a railroad man, the other as a pilot and would-be mechanic. He was spending every spare minute at either Ocala or Hart Field in Jacksonville. If he wasn't busy trying to build flying time, he was helping people rebuild aircraft, which logically led to the purchase of his first "real" airplane, an uncompleted Heath Parasol with a Henderson motorcycle engine. One of the most popular homebuilt designs in the 1920s and 1930s, the Heath project was far from complete, requiring Curtis to test what he had learned by building his first set of wings from scratch. This was also Willie Mae's baptism in airplane building, since she covered and ribstitched the wings. That wouldn't be the last time she performed that duty for Curtis.

"I had worked with a number of different kit plane plans by that time," says Curtis, "and the Heath's were about as good as most. I think working with those different plans had a lot to do with the way I drew my plans later on."

By the time the navy established the Naval Air Station at Jacksonville in 1940, Curtis was ready to leave the railroad. Recognizing his skills and experience the navy hired him as one of their first aircraft inspectors. Specializing mainly in the inspection of welded components, he worked on everything from Stearmans to PBYs.

World War II changed everyone's priorities almost immediately, but Pitts still managed to keep his hand in general aviation. The navy had released Hart Field, an auxiliary field located west of Jacksonville, very early in the war and Pitts began to spend every spare minute there. In the worst way, he wanted to get

into the aviation business, but couldn't afford to give up his navy job, so he took on a partner to help him out. Walter Williams and Curtis worked together and established a flying school at Hart Field, complete with a Cub Coupe ("my pride and joy," says Curtis) and Taylorcrafts for trainers.

As the war pumped money into the economy and the country began to come out of the depression, the Pitts-Williams flying service gained momentum and it looked as if they had made the right decision at the right time. At least it looked that way until a late afternoon tornado tore across the field and their entire business was reduced to scrap. In a matter of seconds, they were out of the aviation business. Curtis' beloved Cub Coupe was found a half mile from the airport, barely recognizable.

The remnants of their business, including the hangar and office rights, were bought by Laurie Yonge who continued to operate the FBO and Curtis continued to flight-instruct part time.

As one of the more popular aviation hangouts, Hart Field, with its 1,800 and 1,200 ft. grass runways and restaurant, was a favorite stop for navy primary students who were always boasting about their latest aerobatic adventures. Pitts listened closely and found himself so drawn to aerobatics he went so far as to travel to Miami to see Mike Murphy flying his Bucker Jungmiester at the All-American Air Maneuvers. Murphy's flying had a tremendous effect on Curtis and he was bitten by the akro bug in a big way.

No longer concerned primarily with the development of the then fledgling corporate aviation market, Pitts' interests gravitated toward such topics as how easily a given airplane would do a loop, or how well it performed when upside down, rather than how long it took to fly from Jacksonville to Miami. Curtis and his friends were more concerned with defying gravity than beating the clock. They were more concerned with discovering what they could do with an airplane than what an airplane could do for them. Names like Tex Rankin and Len Povey, Mike Murphy and others were near legends because of the unbelievable precision they brought to their flying. This was the segment of aviation in which Curtis decided to immerse himself.

Soon, Curtis had had enough of listening to everyone talk about slow rolls and hammerheads, and he decided it was time to take an important step and buy his own aerobatic airplane. On the advice of several local so-called experts, he bought a WACO F-2 with a 100 hp Kinner engine. This, they said, would be a good aerobatic airplane. The advice turned out to be, at the very least, a gross exaggeration, and the airplane turned out to be an aerobatic turkey.

"Of all the airplanes I've owned in my life, that WACO 'F' had to be the sorriest," says Curtis. "It would barely do a roll and you could hardly get high enough to get speed to do a loop. It was just plain sorry."

Having flown a supposedly aerobatic airplane, Curtis was certain there had to be something better, and felt it wouldn't take much to build a machine that would make the WACO F look silly. At that time he had had no experience with some of the better aerobatic airplanes, such as the Bucker Jungmiester, and it is interesting to speculate on what might have happened had one of those been his first aerobatic airplane rather than the WACO. Who knows, he might have left well enough alone and his dissatisfaction might not have overcome his common sense, stopping the first Pitt Special before it was born.

The up side to working for the navy was it exposed Curtis to many highly educated engineering types and taught him the value of having those with the right kind of knowledge available. Although he had only finished high school, he soaked up engineering skills like a sponge, but never got to the point where he wasn't insistent on having someone with the right credentials looking over his shoulder. He wanted to be certain the decimal points were in the right place and the structure was safe. One of these he managed to have checking his decimal points was Dr. Fred Thompson, who would re-surface later in Pitts' life to render similar looking-over-the-shoulder advice.

As the war began to wind down, he began scratching more frantically on notepads, eventually arriving at a design that would be known as the Pitts Special. He knew what he wanted. Now all he had to do was build it.

Number One
The Pitts Special Is Born

Almost from the first flight in the WACO F, Curtis Pitts had been trying to scratch an itch. He knew what aerobatics took: it took the ability to roll, the ability to climb, the ability to move quickly from one attitude to another. And size is one of the most important parts of the equation. The smaller the airplane, the less power it needs. The shorter the wings, the faster it rolls. The lighter the airplane, the faster it climbs on a given engine and wing size. None of these concepts were unheard of, but they had never been combined into a single aerobatic package before. The name of the aerobatic game, although no one had said it before, was pack as much punch in as small and light a package as possible.

As is always the case with airplanes, engines drive the design, and in the case of earlier aerobatic airplanes, they had

The rigid landing gear and balloon tires of the Number One Pitts look out of place, but worked fine.

The original Pitts Special in the hangar at Hart Field, Jacksonville, Florida, prior to the airplane's first flight in 1945.

Number One on its way to the airport for the first time. It hasn't been painted yet in this picture.

to be big enough to carry the engines available at the time. With even the lowest horsepower radial engines of the day weighing 400 and 500 lb. it was very difficult to design a small, high-performance airplane around them. However, by the time the war began and the aerobatic visions dancing around in Pitts' head were beginning to take shape, the variety of lightweight, horizontally opposed engines was expanding greatly.

Small four-cylinder Lycomings, Franklins and Continentals of 55 to 100 hp were being used in increasing number by all the major manufacturers. None of these engines weighed much over 175 lb., which totally changed the ability to design a little biplane that could carry a man into attitudes and angles never dreamed possible.

Since Pitts was designing and building on a workingman's wages, cost was a

The size comparison between the young boy and the Number One Pitts gives some idea of the airplane's dimensions.

major factor and he couldn't drop a lot of cash on a counter and walk out with a new engine tucked under his arm. That's where the tornado paid him back a portion of what it took. One of his school T-Crafts had been reduced to junk, but the engine was unharmed. A few hours with the tool box liberated the 55 hp Lycoming 0-145 from the wreckage and the little airplane project began to move ahead in earnest.

There was never any thought given to anything other than a biplane layout,

Lauri Yonge's son in front of Number One. Yonge was a major figure in Florida aviation and ran the FBO where Curtis Pitts gave flight instruction.

since that combination of wires and struts always creates a wire-braced cage that gives more strength for less weight than any other structural approach. Pitts knew this was not necessarily the most aerodynamically clean arrangement, but he wasn't looking for speed, he was looking for low weight and high lift, a combination always favored by biplanes.

Since pilots can't be scaled down, the minimum size of the airplane was dictated by the basic Curtis Pitts-sized pilot, which further determined the span and chord of the wing, and therefore the wing area.

The final wingspan came out to be a miniscule 16 ft., 7 in. for the longer top wing, with a total area of approximately 90 sq.-ft. This certainly made it one of the smallest biplanes built to that time.

The top wing was swept back 6½ deg. for several reasons, the most obvious

being it made it much easier to get into the cockpit. Actually, it could be said it made it *possible* to get into the cockpit. In such a tiny airplane, the pilot has to sit well forward in the fuselage. Unfortunately, that is also about where the wings have to be. Had the top wing been straight and in the proper position in relationship to the center of gravity, Pitts would have had to saw a hole in the middle of it just to get into the cockpit. Moving it forward gave access, and sweeping the wings back put the center of lift back where it should be.

"Yes, I swept that top wing for CG and access reasons, but some of the engineers at the navy lab where I was working said I'd also pick up a little stability, so that was another consideration," says Pitts.

Although it wasn't a driving factor in the let's-see-how-fast-we-can-make-an-airplane-roll equation, the top wing

Number One pulls a hard bank over the water. The replica was built in 1990 in only five months and is now part of the Pitts display in the EAA Museum in Oshkosh, Wisconsin.

sweep also gave him a snap-rolling ability that redefined the word awesome. Undoubtably, the first time he snapped the airplane, he thought his eyeballs had been twisted into knots because of the speed at which the horizon whipped around him.

The wings of the Pitts Special are unique, if for no other reason than the top one is a single piece. Traditionally, biplanes' top wings are either hinged at the middle and made in two panels, or joined to a stationary center section. Both methods, however, require many more fittings and pieces which increase the weight and complexity of the pro-

It's hard to believe the first Pitts had only 55 hp. Note the headrest on the turtle deck.

ject. Curtis was looking for the simplest way to have a strong, lightweight wing and the best way to do that was to join the two spars in the middle with a pair of splice blocks, thereby making them a single unit. That's one nice thing about building baby biplanes; even the big pieces are small.

Since he knew weight was his biggest enemy, especially with only fifty-five horses hiding in the nose, Curtis went out of his way to eliminate unnecessary parts and structure. The result was the wing structure on the original airplane and on every Pitts Special since, which often surprises folks with its model-airplane appearance. The ribs on the Number One airplane, for instance, were fragile trusses of ¼ sq-in spruce, and are

still that size. That means the ribs are built up of pieces significantly smaller than the average pencil. The spars in that first airplane measured only ⅝ in. thick, about the dimension of a cheap paperback. But, he couldn't afford the weight, so he worked with his slide rule and engineering friends to determine exactly how much wood was actually needed to withstand the rigors of aerobatic flight. When he was finished, that first wing probably didn't weigh much more than the seat cushions in his old WACO F.

When Curtis put the full-span ailerons on the two bottom wings, he was doing it partially because of convenience, since it was simple to do it that way. But he also wanted the airplane to have as much roll rate as possible, so he gave it as much aileron as possible. On the first flight, he found he had accomplished what he wanted. "I knew those big aile-

rons would help it roll, but it wasn't until that first flight I realized just how well they would work," recalls Curtis.

Essentially, the skeleton of the Pitts Special, as it would eventually be called, was totally unremarkable except for its size and weight. The steel-tube fuselage was less than 18 in. wide inside the longerons, barely allowing Curtis, who appears quite thin in old photographs, to squeeze into the tiny cockpit. The tail too was a steel tube unit with bent-up sheet steel ribs only 0.025 in. thick.

The main landing gear was unique, considering it was totally rigid with the only shock absorption being supplied by the 700x4 balloon tires off an E-2 Cub. It was a separate unit that ran, in an inverted U shape, from axle to axle and was then bolted to the bottom of the fuselage. There were no brace tubes between the wheels, which means it could take only minor side loads with-

16

out bending the tubes. Small streamlined wires from the axles up to the fuselage kept the gear from spreading out on impact. Although the concepts of the original airplane have continued almost unchanged through the entire Pitts Special saga, the rigid gear was unique to old Number One. All the rest had shock-absorbing, bungee-supported landing gear.

The first few flights of the airplane were made without the benefit of streamlined wires between the wings. Instead, stranded rudder cable was clamped in place, which did the job admirably but contributed considerably to the airplanes' drag. Looking at those tiny wire ropes between the wings from this vantage point in history makes the airplane look almost crude, but we must remember that streamlined wires have always been expensive and Curtis was building an airplane between the cracks of raising a family. The cables were replaced as soon as the money was available.

In a bargain-basement project like the Special, much of the design is determined by what is available in the scrap heap. Besides yielding the E-2 Cub tires and brakes it also produced the nose bowl. Again, it was one of the school's recently deceased BL-65 Taylorcrafts that gave the Pitts Special its distinctive nose with its grillework.

Curtis started cutting metal on the airplane in late 1944, actually doing most of the building during 1945. He was working at home in his garage, although Willie Mae covered all the control surfaces on their kitchen table. The airplane was covered in Grade A cotton and finished in dope, with the final color being off-white with maroon trim.

After spraying on the last coat of dope and pronouncing the project finished, he loaded it on the back of a big old flatbed and hauled it out to the airport for final assembly. When the airplane was ready for flight, Curtis called the local CAA representative (this was prior

Minimal would be one way to describe old Number One's panel. Take a look at the round tube cabane struts.

Next page
The 1945, Number One Pitts replica displays the wing and tail planform identified to Pitts Specials to this day.

45 years separates these two photos. Curtis Pitts recreates the photo taken shortly after he first flew Number One.

to the FAA) to inspect it before it made its first flight. The CAA sent over two inspectors, one from maintenance and one from flight-test, and the second they saw the airplane they cast an incredulous look first at the airplane, then at Curtis. They shook their heads and immediately started talking about how quick and unstable it would be. They didn't realize those were also some of the design goals.

As Curtis tells it, "The CAA guys came out there three different times and just couldn't bring themselves to sign the paperwork on it the first two times. Its size seemed to make them too nervous."

What the CAA didn't know was that between visits two and three, Curtis ran

A T-Craft donated the nose cowl and grille work on the original Number One.

out of patience and, having taxied up and down the runway for several days, had had about all he could take. It was time to fly the airplane and see what it would do. In fact, by the time the CAA came back the third time with the completed paperwork in hand, he already had twenty minutes on the airplane (a fact he didn't bother sharing with the inspector).

"As I sat on the end of the runway in that little airplane I don't think I was scared, but I was certainly nervous. When I brought the power up, it accelerated far faster than anything I had ever flown and bounced about twice before leaping into the air. It seemed so sensitive in pitch that I was either diving

or climbing for the entire ten minutes I was up, and I was really nervous at that point. In fact, I was convinced we'd gotten the CG all screwed up," remembers Curtis.

Back on the ground after that first flight, they adjusted the stabilizer angle of incidence, the tail heaviness disappeared and Curtis was in love. Even before the airplane had gotten its official paperwork from the CAA, he knew he had accomplished what he wanted to do.

After the CAA gave its approval, Curtis' close friend Phil Quigley decided to play a practical joke on him. He had Curtis convinced the CAA had seen him fly the airplane illegally and were getting ready to cite him for violating federal regulations.

"I didn't sleep for two nights," laughs Curtis. "I was still instructing for a living

and I needed the money, so I couldn't afford to lose my ticket. I don't think Phil knew the effect the joke would have on me."

Incidentally, for a short time they called the airplane Jeep because of its tiny size. Then they realized Art Chester had a racer by that name, and it became the Pitts Special.

Curtis continued refining the airplane, replacing the 55 hp Lycoming after a short time when a 90 hp Franklin became available. With that giant jump in horsepower, the airplane, which had been impressive with the original engine, became an absolute skyrocket and a local legend.

With the new engine, the plane would zip down the runway and into the air in a show of performance local pilots had seen only from the military hardware of the recently finished big shoot-out. They

20

were lined up to have a go at the airplane; one of them was Courtenay Hunt. He remembers his first flight: "On the ground I couldn't believe how small it looked, but, as soon as I was in the air I wondered what the hell I had gotten ahold of."

Hunt remembers sweating the landing, but had no trouble with the airplane at all. He said many local pilots lined up to fly it, and none had any problems with it.

According to Hunt, the airplane fit right into the local atmosphere of fun and friendship which seemed to center around Curtis and Willie Mae, who ran the restaurant.

"In the first place, Willie Mae was a really good cook, and Curtis was hard not to like. As an instructor, he was excellent, because he was always encouraging his students and almost never

yelled. He made the airport a great place to be."

As much fun as the airplane was to fly, it wasn't without its problems, one of which was the rigid landing gear. On one particularly memorable day Curtis himself hit the rough sod of a friend's strip a little harder than he had intended and bounced mightily. What he didn't know was that he had impaled the axles into the ground, leaving both wheels on the spot, while he bounced merrily back up into the air, having no idea that all he had under him were the pointed Vs of the now-wheelless landing gear. Naturally, as the airplane came down out of the bounce, it behaved exactly as the laws of physics dictated: the gear punched holes in the ground and stopped instantly. The rest of the airplane proceeded to rotate about those two fixed points, depositing Curtis

Number One in its final configuration with 90 hp Franklin engine.

and the airplane on their collective heads in something under a nanosecond. The airplane had barely stopped its backward, upside-down slide when Curtis hastily scrambled out from under it, expecting fire at any minute and still not understanding exactly what had happened. That became clear, however, when one of the landing gear tires caught up and bounced happily past him, its inertia still unspent.

Relatively minor crashes of one-off designs always give the builder an opportunity to change things. In this case, old photos show the round cabane struts of the original version gave way to the more typical streamlined tubing almost always seen in this application.

After flying the airplane for several years, Curtis sold it to a crop duster from North Carolina. Both Hunt and Pitts remember him as being so deaf, he couldn't hear an engine running if he was right next to it. But, he climbed into the airplane and took it north.

A number of the duster pilots flew the airplane, but Jack Reynolds let his exuberance and good judgment get out of hand. The story varies, but the airplane was too low and the engine quit while inverted. By the time Reynolds realized it wasn't running, it was too late. He got it right side up but mushed it into the ground. Hard! As the Special plowed its way through the dirt, it rolled itself into a ball that, as Pitts says, "would fit into a number 10 washtub." Miraculously, the pilot limped away in much better shape than he should have been.

The Special wasn't so lucky. There was practically nothing salvageable, so it was left to rot until finally being ingloriously carted off to the dump, an ignoble end for such a happy little airster that was the seed for so many legends to follow.

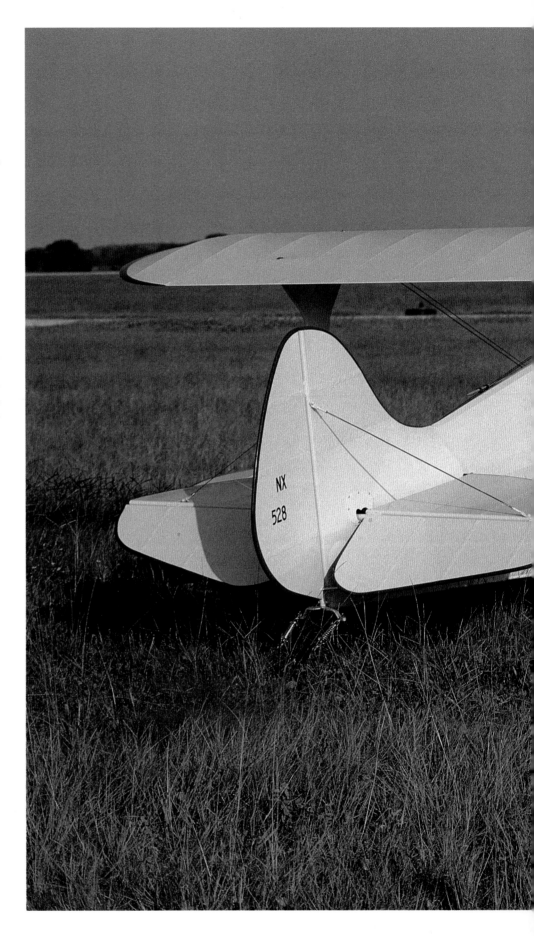

The vertical fin and turtle deck of Number One are slightly different, but the plane's general shape and wings are definitely Pitts Special.

Famous Planes, Famous Names
The Air-Show Game

By the time the war was over, airplanes had a definite identity crisis: For half a decade, their primary purpose had been to reduce large populations to cinders or to train and support those who could. By the same token, however, the United States had a severe case of the good-time blues; they were ready for smiles and laughter. They wanted to be entertained, and the pilots and airplanes returning from the shooting were all too happy to oblige. Aerial entertainment in the form of air shows and races began to pop up all over the country, and there were a few folks who thought the little airplane Curtis Pitts had crafted would offer a nice change of pace from the burly Stearmans and snarling Mustangs.

Pitts had always had his eye on aerobatics, but it wasn't until Carl Stengel from Gainesville, Florida, approached

Phil Quigley was an integral part of the early Pitts legacy. He's seen here flying the Number Two airplane before it became Betty Skelton's Little Stinker. *Betty Skelton flew part of her first airshow season with the airplane in this paint scheme.*
D. McGuire

Skelton and Little Stinker *were loved by the press and publicity photos abound. Here she plays with* Little Tinker, *named for Air Force General Tinker.*

him that he began to think of the little Special as being a commercially viable airplane. Stengel was the fixed base operator at Gainesville, and he felt certain there was a ready market for the frantic little bumblebee with both sport and air-show pilots. He talked Curtis and Willie Mae into moving to Gainesville, where he and Pitts would produce ten airplanes and offer them for sale.

Undoubtedly, Curtis and Stengel looked around at 1946 aviation and thought the unbelievable boom they were seeing was the real thing. Every manufacturer with as much as a welding torch was cranking out airplanes like a cookie factory. In fact, during 1946 alone, over 35,000 airplanes were built. Unfortunately, almost none of them were sold and the next year, a tenth that many planes were built. The year after that, many of the factories either shut off their lights, or reduced their staff to those needed to sweep the floors and do a little rib-stitching. The superboom had lasted less than two years.

As is often the case with entrepreneurs, the dream was easier than the reality: No sooner had Curtis and Willie Mae unpacked their belongings in their new surroundings than Stengel found himself in financial trouble. He was forced to sell his business, and the successful bidder turned out to be a couple who'd recently moved from Jacksonville—the Pitts. Curtis may have thought he was coming to Gainesville to build

little airplanes, but he wound up building a business.

Part of his newly purchased inventory was a redesigned Pitts Special fuselage sitting on its landing gear. Although they hadn't gotten very far on building that second airplane, he had spent a lot of time on the drawing board redesigning the airplane. Dr. Fred Thompson, his old engineering mentor from NAS Jacksonville, was by then part of the aeronautical engineering department at the University of Florida at Gainesville and Curtis once again called on him for help. Although Curtis had already redesigned and re-engineered the airplane, he wanted a professional to look over his work so there were no mistakes.

The Number Two airplane benefited greatly from Curtis' experience with the first machine. Among other things, most of the fittings were greatly simplified by replacing the many thin and bent-up pieces with flat fittings of slightly thicker gauge but of much simpler design. It was also slightly bigger than the first plane, the result of Curtis' friends remarking how hard it was to put a normal-sized pilot into a less-than-normal-sized cockpit. Phil Quigley and other shorter men found the airplane adequate, but those long of leg were folded up like jackknives, with no place to put their elbows. As a concession, it was made slightly wider. Even so, it was still a tight fit for many.

Betty Skelton and Little Stinker *combine to produce what is undoubtably the most famous Pitts picture ever taken.*
B. Frankmans

Although the 90 hp Franklin had made the original airplane into a rocketship, Curtis felt a little more wing was in order to give better climb and glide performance. The second airplane was fitted with panels 6 in. longer.

Looking back, the landing gear on Number One was something to behold. Since it was absolutely stiff, it allowed absolutely no room for error and couldn't compensate for a rough runway or a rough pilot.

On Number Two, the landing gear was changed to a conventional tripod-type V pivoted on the upper portions of the V with bungees wrapped around the center portion, giving the airplane a badly needed shock-absorbing landing-gear system. As it happens, most Pitts drivers to this day think their gear is practically solid, because the bungees are so stiff. Until they fly a rigid gear, they don't know what stiff really means.

Another change which has remained with the airplane to this day is the turtle deck shape. The original turtle deck was composed of twenty-two thin stringers formed into a headrest at the front and the characteristic Pitts Special inverted V at the back. This was not only a pain to build, but broke the lines of the airplane. On Number Two, the bulkhead directly behind the pilot's head was formed into a slightly curved inverted V (just like the back one), and the entire turtle deck was wrapped with thin sheet aluminum. Forty years later, the Pitts Specials being built use the exact same technique and shape.

Equally as important as the airframe changes was the C-85-F5 Continental hung in the nose. This lightweight, fuel-injected beauty fit nicely under the cowling while keeping the weight to the absolute minimum. Essentially the same engine being used in the contemporary Ercoupes and Luscombes, the engine pulled the Number Two Special to aero-

batic heights most pilots couldn't even imagine.

It's important to remember that although this plane had only 85 hp, as compared to the 180 or 200 hp pilots came to expect in later years, the airplane was really light. In fact, it weighed in the area of 560 lb. empty, as compared to 750–800 lb. for later models. The weight difference may not sound like much, but it is significant. In fact, the Number Two airplane complete with pilot and fuel weighed no more than a later S-1T empty. The airplane's naturally light weight combined with its

small size to make it a machine that could easily ignore gravity.

Structurally and aerodynamically, the second Pitts Special is an especially important airplane because it set the patterns for all Specials to follow. In that respect it may even be more important than the first, since it represents the first major step into what was to become the present-day Pitts Special. It was given bigger engines and a longer fuselage and cockpit, but the basics remained. Until the all-out competition machines of the mid-1960s, when the semi-symmetrical M-6 airfoil was

At 5 ft., 3 in., 1951 womans national champion Caro Bayley makes the third Pitts Special look large. Airplane had 125 hp and was known as Black Magic. C. Bosca

changed in favor of a Pitts-brewed symmetrical section and an extra set of ailerons was added, the airplanes remained essentially the same as Number Two.

The second airplane had been completed and Curtis and his local friends were flying it when Jess Bristow, noted air-show pilot and entrepreneur from

Pert Caro Bayley, 28-year-old aviatrix from Springfield, Ohio, was scheduled to make an official attempt to break the women's international altitude record for light planes earlier this week. She is also the only woman flier entered in the National Aerobatic Championship Contest, in which she will match her pilot skill against the cream of the country's top airmen. The mark Miss Bayley was to shoot for was 24,504 feet, set by Anna Luisa Branger, of the Venezuelan Embassy in Washington, D. C. Last January the attractive Miss Bayley climbed to 27,510 feet in an unofficial attempt. Caro is an accomplished precision flier in her own right and promises to give the boys plenty of opposition in the aerobatic contest. During World War II Caro flew with the WASPS, flying various type military aircraft from heavy Thunderbolt fighters to multi-engine bombers. Since the war she's been campaigning as a professional aerobatic performer in all major air shows throughout the country. Her brilliant career in aviation has also included flight instructing and a short period as an airline stewardess. However, handling an airplane is only one of her many achievements. Caro is a talented artist and has ambitions of someday becoming a fashion designer. In the aerobatic event she'll fly her new Pitts Special, desined and built by Curtiss Pitts in Gainesville, Florida, which is powered by a 125 horsepower engine. For the altitude record attempt Caro flew a Piper Super Cub, also equipped with a 125 horsepower engine, which is a product of the Piper Aircraft Company, Lockhaven, Pa. A sealed barograph, regulated by the Bureau of Standards, and special oxygen equipment was installed in the plane. The altitude flight was sanctioned by the National Aeronautic Association, the U. S. affiliate of the Federation Aeronautique Internationale.

Caro Bayley's exploits weren't limited to the Pitts Special, as this 1952 clipping shows. Photo is in front of Stengel Flying Service at Gainesville, Florida, where Pitts based his business from 1946 to the late 1950s. This hangar was home for all of the early Pitts designs. F. M. Rogers

North Carolina, saw the airplane and immediately recognized its air-show potential. Even though Bristow couldn't even begin to fit into the airplane, he bought it almost immediately and contracted with Phil Quigley to fly it as part of his air shows.

Quigley flew the airplane for one season but the relationship never proved successful so, when a young lady from Tampa, Florida, made an offer to buy the airplane from Bristow, he saw no reason to turn it down since he himself couldn't fly the plane. His sale of the Number Two Pitts Special to young Betty Skelton proved to be one of the more important moves in postwar sport aviation.

Betty Skelton was a highly unusual person in many ways, not the least of which was her boundless energy and her determination to do things exciting and do them right. Combine those qualities with striking beauty, a petite frame and a natural drive toward making the best of her own talents and one wonders why Hollywood hasn't picked up on her story.

Skelton got into aviation at a time when a woman's place was still considered to be in the kitchen. Amelia Earhart and Pancho Barnes were the exceptions. However, she could hardly have avoided the aviation bug, since her childhood home was located directly in the middle of NAS Pensacola's aerobatic training area. If she looked up from play, there was always a bright yellow N3N or N2S cavorting above her.

The aeronautical environment affected her entire family. Her father, who worked for the railroad, began taking lessons from a navy ensign in a 40 hp Taylorcraft, as did her mother. At the time, Betty was twelve years old and remembers it well: "The ensign gave me

lessons too," she giggles a little, "and one Sunday he soloed me even though I was only twelve years old. Dad made me promise not to tell Mom. We couldn't tell anyone."

She hung around the airport as the stereotypical airport kid, bumming rides and flight time in all the CAP (Civil Air Patrol) flights the war generated. Then, on her sixteenth birthday, she soloed that same T-Craft, NC22203, which was an anticlimax only to her, since she was still a local celebrity and one of the, if not the, youngest female pilots in the country.

Skelton's father moved the family to Tampa to open a Civilian Pilot Training (CPT) school to provide primary training to military cadets. She continued flying, but the day after graduation from high school she went to work for Eastern Airlines as a night agent, while continuing to flight-instruct in her father's school during the day.

Skelton recalls, "I wasn't a very good instructor. I couldn't remember a time when I didn't know how to fly. It was just something that seemed natural and I did it without thinking. But, I don't think I was able to tell others how to do it very well."

Her goal to be a WASP ferry pilot was frustrated because just a month before reaching the required 18½-year-old age limit, they dropped the WASP program.

Skelton gave a passing thought to being a flight attendant for Eastern, but one stint at substituting for a sick attendant on the Tampa-Miami run convinced her she should look elsewhere in aviation.

The elsewhere actually came looking for her. The local chamber of commerce was putting together an air show and thought it would be unique if "that young girl out at the airport" would fly in it. At the time she had done no aerobatics, but there was a month before the show would go on. Her dad prevailed on Clem Whittenbeck, a noted aerobatic pilot of the time, to give his daughter some instruction in one of the school's PT-19s.

"Clem taught me how to do a roll and a loop," she remembers. "I looped and rolled back and forth across the field, never getting below 2,000 feet."

One of the spectators was a promoter for the Lake Wales, Florida, air show and asked her if she'd perform there, and, as the saying goes, history knows the rest.

She considers her first professional show to be in Jacksonville in 1946, which coincidentally was also the first showing for a new group calling themselves The Blue Angels. She was nineteen years old at the time.

After the spars in five different PT-19s showed signs of coming apart because of her using them in air shows, Skelton's dad decided it was time for something a little sturdier. He found a derelict Great Lakes which he and Whittenbeck rebuilt, hanging a 160 hp Kinner in the nose. The Kinner was not known as one of the world's great engines, but it was all they could afford.

Skelton could, and should, write a book about her experiences in flying that Great Lakes all over the United States, from Pennsylvania to California, and the problems she had with it. Like the time she fell asleep on a cross-country and awoke not knowing how much time had elapsed because her clock had stopped. That was important, since she had no gas gauge. Chalk up another forced landing, when she ran

out of fuel short of Tucson. Or the time the center section caved in at the bottom of an outside loop. The FAA grounded the airplane, but she had to get it home, so she sneaked off and waddled all the way home from North Carolina near stall speed.

Skelton could also tell about the time she had been sponsored to break Jackie Cochran's speed record and Woody Edmundsen loaned her his P-51. She had never even sat in one. With four hours of T-6 time and one takeoff and landing in Edmundsen's airplane (who was as scared as Skelton was), she went roaring off across Tampa Bay. She had the record made, at over 420 mph, when, on the fourth pass, the engine blew up. It didn't simply quit. It blew up and was trailing smoke, fire and pieces across the water. She pulled up and had her hand on the canopy release, when she looked down at Tampa Bay and decided to dead stick it into McDill Air Force Base. Skelton remembered she didn't know how to swim. Later, Edmundsen said he wished she had bailed out, since he had it insured.

Skelton's love affair with Pitts Specials began as she was coming into the pattern at the 1947 air maneuvers in her Lakes. She looked down, saw this tiny airplane surrounded by people and even from the air, she knew she was in love. The second she was out of her airplane, she rushed over and pushed her way through the crowd to look at what she thought was the most perfectly proportioned plane she had ever seen.

"I love small things. Tiny things. Maybe it's because of my size, but the Pitts just looked like it would fit me exactly right," she remembers with fondness.

Her request to sit in the airplane resulted in a chauvinistic rebuff from the three gentlemen around the airplane, one of whom was Phil Quigley and another who most certainly was Curtis Pitts. In later years they would laugh about it. At the time, she stomped away, determined she'd own that airplane someday.

Although Curtis rejected her initial attempts to buy the airplane, Skelton kept track of the Pitts until a year or so later when she heard it might be for sale. Calling Bristow, she settled on the unseemingly high price of $3,000, and agreed to pick up the plane at Gainesville.

Recalls Skelton, "$3,000 doesn't sound high now, but it might as well have been a million to me. In my family that much money was simply unheard of."

Her first flight in the airplane was also the ferry flight from Gainesville to Tampa. She remembers her first takeoff well. "It was the most amazing experience. I opened the throttle and was in the air, just that quickly. I looked out at those tiny wings, felt the astounding sensitivity and couldn't believe I was there."

Skelton also remembers her first landing in the airplane. The press had been alerted to her arrival and stood with her friends and father in front of the hangar awaiting her return. They all got a front-row seat to see a young lady learning what generations of pilots would eventually learn about the Pitts—the landing is the tough part. Not aware that rudder effectiveness would disappear long before the airplane stopped rolling, she didn't have her feet positioned on the brakes and did the cutest little ground loop, right there in front of God and what seemed like half of northern Florida.

"I was so terribly embarrassed. I hate it when I do something wrong, and that was right in front of everybody," says Skelton.

Almost from that day on she and the airplane were melded together as one. In fact, her air-show schedule had gotten so hectic, it was well into the season before she got the time to repaint the airplane to her own color scheme.

She wanted to get rid of the huge original number, NX864019, and applied to the CAA for the smallest number they had. N22E was what they came up with. Looking for a name for the airplane, she remembered when she was the young kid hanging around the airport, her dad and others called her "Little Stinker," and that's what she decided to name her plane. The feminine-looking skunk painted on the turtle deck completed the image.

It's difficult to paint a concise picture of Skelton's activities after acquiring the Number Two Pitts because they were so diverse and intense. In a matter of a few years, she became a national hero by virtue of her high public profile and pressworthy activities. She was cute, she was exciting, she was just what the

Dean Case and his daughter Joyce posing with the first Case-Pitts, circa 1959. Due to lack of plans, Case made many modifications including aluminum ribs. Case/James

press loved. She was also damned good at what she did.

Besides becoming female aerobatic champion three times, Skelton championed the cause of aviation wherever she could, going on endless speaking engagements and flying at literally hundreds of air shows.

Possibly her proudest achievement was in representing the United States in the International Air Pageant in Gatwick, England. She flew the *Little*

Stinker to Newark, New Jersey, where it was dismantled and lovingly placed in a shipping crate before being put onboard the *Queen Mary*. Dismantled is too easy a word for the huge task she and the local mechanic tackled. There were no manuals for the airplane, so they had to undo the flying wires and count the number of turns to be sure the airplane would be in rig when it was reassembled. The sponsor, the *London Daily Express*, wouldn't pay to send a mechanic along with the airplane, so Skelton had to hover over the English mechanics, making sure everything went back together properly.

"On the way over, we had some awful weather and I couldn't stop worrying about my airplane," she remembers.

"The weather was so bad, that at one point the only people at dinner were me and the crew. I pestered the captain until he let me go down in the hold and convince myself the crate was securely tied down. Then, I felt better."

Once in England, she began to enjoy herself. "The English were fantastic!" Skelton remarks. "They had never seen anything as small or as nimble as the *Stinker* was, especially being flown by a woman. They treated me great!"

Part of the trip included a flight over ninety miles of water with a twin-engine Royal Air Force (RAF) AVRO Anson for escort. The RAF had sent the Anson out, complete with survival rafts, ditching gear and everything else needed to help her in the event she had to ditch in the

Joyce Case poses with Joy's Toy, *the first Pitts her father built for her.* Case/James

water. They were missing only one thing: someone in the Anson to throw the gear out. The RAF pilot was the only crew member on board!

On the return trip, Skelton simply taped over all the engine instruments so, if something started to go wrong, she wouldn't know about it. Her escort this time was a new found friend in a Spitfire that S-turned around her with his gear down for the entire overwater portion. Obviously, the little Continental kept running.

During the late 1940s and early 1950s, Betty Skelton and *Little Stinker* wrote the first serious chapter in what was to become a larger book in the history of aerobatics. More than that, they popularized the image of a small, agile biplane

as something that gave pilots the freedom to do anything they wished.

It's difficult to calculate what impact Skelton and her airplane had on future generations of sport pilots because of her high visibility and exciting adventures. When the airplane was put into model-airplane kit form, it instantly touched the lives of many young modelers who were mesmerized watching it whine away on the end of a pair of wires leading back to a handle in their hand. Twenty years later, the same modelers became the basis for the surge in sport aviation and aerobatics which occurred in the mid-1960s.

Aside from Curtis Pitts himself, the two most important names in the Pitts Special legacy would have to be first

Betty Skelton, and later Bob Herendeen of whom we will hear more about in later chapters.

Probably because of its small size, the Pitts became the airplane of choice for many female aerobatic pilots in the 1950s. Following shortly on Betty Skelton's heels came Caro Bayley. Even more petite than Skelton, Bayley, a former WASP pilot, could look back at hundreds of hours in B-25s, AT-11s and even P-47s during her tour in the military. After the war she cut her teeth in the air-show business flying a clipped

Cub for Jess Bristow, and later booked herself in the Bradley Special, a Cub-type specialty akro machine.

Bayley had been competing against Skelton, usually coming in hard on her tail, while Skelton was flying her Great Lakes and Bayley her Bradley Special. When Skelton switched to the Pitts, the game was all over.

Then, at one of the All-Girl Air Show performances, Phil Quigley and Curtis Pitts had the Number Two airplane with them. A little arm-twisting on Bayley's part put her in the little plane's cockpit and then in the air.

"That was it," Bayley recalls. "I knew that was the airplane for me."

Curtis had a third airplane already well under way with a 125 Lycoming in it and Bayley said, in no uncertain terms, that she wanted it. Her father flew down from Springfield, Illinois, to look at the airplane and evaluate this man, Curtis Pitts. Satisfied, he gave his daughter the go-ahead. "When I gave Curtis a check for $3,000, all I could think was how messed up the economy was," says Bayley.

Bayley moved to Gainesville and camped on Quigley and Pitts' doorstep until they finished the airplane. Laughingly and lovingly, she remembers them as being "ornery and lazy."

When the airplane was about ready to go, Quigley and Bayley got into a friendly argument about who was going to get to fly it first. Quigley settled the argument by laying down his tools and saying, "you can fly it okay, but it will have to be like this . . .," and he walked away from the unfinished airplane.

The airplane was painted a gloss black and the P. T. Barnum of air shows, Bill Sweet, dubbed it *Black Magic*. Bayley competed and air-showed the plane until late 1951, when she got married. She wasn't flying the airplane much and her father told her she should pickle it and put it in storage. Instead, she sold it. Today, she tells her kids and grandkids, "always listen to your father. . . ."

The new owner, Frank Gibson, had a fuel-injection line break in-flight and a fire broke out almost immediately. He got the airplane down in a field, safely, but had to stand by and watch it turn to ashes. This left *Little Stinker* the sole survivor of the original batch of Pitts Specials.

Air shows nearly went out of business in the early 1950s because of a tragic accident reportedly involving a transient pilot slow-rolling into the crowd. When the interest in air shows died, so did interest in the little airplane. Just when it looked as if the Special was about to become an air-show legend, Curtis' phone simply stopped ringing. Still, there were those who wanted to build the plane, one of whom was Dean Case of Wichita, Kansas.

Case, a corporate pilot and mechanic, had a daughter named Joyce whom he had groomed, as a teenager, to solo on her sixteenth birthday in 1953. She did solo on that date, but the aviation transfusion wasn't as strong as the more normal interests of a sixteen-year-old girl, and it would be several more years before she would begin to develop as a pilot.

In 1959 Joyce Case took the Warner-powered Great Lakes her father had rebuilt only a month before and took off for the Antique Airplane Association's aerobatic contest in Ottumwa, Iowa. The fact that she had been flying aerobatics for only a month didn't seem to matter, since she won the women's division easily. She was also the only woman contestant. Hal Krier won the overall contest, but Case met several other individuals who would later go on to make their own marks on the world. One was a young man named Charlie Hillard, and the other a tall, skinny drink of water named Richard Bach.

Even in 1959 everyone recognized the problems of doing serious aerobatics in the Great Lakes, because it was too big, too draggy and too underpowered. So, it was understandable Dean Case would remember *Little Stinker* and *Black Magic* and how well they performed. He wanted to build a Pitts for Joyce, but plans were simply not available. According to Curtis Pitts, Case bought an incomplete Pitts project from Bill Williams of Tulsa, Oklahoma. Curtis had given Williams the few scraps of paper he had with miscellaneous drawings for the airplane on them.

"There wasn't nearly enough information to build an airplane," says Curtis. "When Dean got the project, he had to figure it out for himself and do a lot of design work on his own."

The first Dean Case-built airplane was about half Case, half Pitts. At this time, the only Pitts planes that had flown were Number One, *Little Stinker* and *Black Magic*. For that reason, the first Dean Case airplane, N11JC, called *Joy's Toy*, was as much a Case Special as a Pitts Special. It did, however, look like an S-1C, flatwing Pitts.

The second airplane he built for Joyce was again modified and was slightly larger. Named *Black Jack*, it was one of the first to mount the 180 hp Lycoming and carried the number N12JJ.

By the mid-1960s Joyce Case had won the national championship and was getting serious about winning. And the competition was getting tough. Curtis Pitts was making the roundwings available, so Dean Case bought a set for the third airplane he built for Joyce. *Joy's Toy Too* was this creation's name, and N12J was the number.

Not as much a darling of the press as Betty Skelton, although certainly possessed of all the right aeronautical attributes, Joyce Case nonetheless made a solid name for herself. During the rebirth of sport aviation and the forming of the first solid foundation for modern competition aerobatics, Case was the woman to beat. Her father should also be credited with helping in building that foundation, if only because he singlehandedly built at least five Pitts Specials, which greatly increased the world's population of competition biplanes.

In the early days of aerobatics, it seemed as if women blazed the path for the Pitts, while men stuck to the Great Lakes, and later the modified Chipmunks. Many of these women who made the airplane visible never made a name for themselves, and others never got their share of credit. Another of the early pioneers was Mary Aiken of Wichita, Kansas, who also used a Dean Case Pitts to do battle in the competitive arena. One of those she frequently battled was Mary Gaffaney from Miami, Florida. Gaffaney would go on to do her country proud.

The Other Pitts
They Weren't All Biplanes

In the world of aviation, the name Pitts automatically conjures up the image of a twisting, tumbling biplane, when the reality is that by 1950, there were as many Pitts monoplanes as there were Pitts biplanes.

Almost as soon as Curtis Pitts moved to Gainesville, Florida, he opened a mechanic school to take advantage of all the GI Bill training available to returning veterans. At the same time, he found himself neck deep in both air racing and building air-show machines for various performers, including Skelton and Bayley.

Although his school kept him busy, he still found time to design and build several Goodyear, now known as Formula 1, racers. The first, sometimes known as No. 27 *Pellet*, was to be campaigned around the country by Bud Heisel, a former navy Corsair driver. *Pellet* featured an all-wood, low wing and a traditional four-longeron tubing fuselage.

Air racing is a risky business, and the races at San Diego proved that. During

Number One and Little Monster *represent the extreme ends of the aircraft design spectrum and were both built in the same five-year period, 1945–1950.*

The least-known Pitts of them all was Little Monster, *the Goodyear racer Curtis Pitts campaigned in the early 1950s. This is the original, restored airplane.*

an early lap, Heisel got caught in the prop wash of several other aircraft, flipping him onto his back. Before he could get the airplane righted, he hit the ground and was killed instantly. The new National Race Pilots Association (NRPA) president, Art Chester, one of the biggest names in air racing, stopped the races until they had a meeting with the pilots to discuss safety. Then, in the very next heat, Chester was killed in exactly the same manner and in exactly the same spot.

Heisel's death saddened Curtis greatly and took all the heart out of him for some time. He pushed the partially completed second Formula 1 racer, dubbed *Little Monster*, into the back of the hangar and forgot about it while he concentrated on other projects.

One of these other projects came in courtesy of his old friend Jess Bristow. At that time most air-show performers were pushing around WACOs or, more likely, surplus Stearman trainers, most of which had received a performance transplant, courtesy of an equally surplus BT–13 Vultee. The entire power section, 450 hp R–985 Pratt and Whitney, cowling, mount and everything was bolted on to give the staid old Stearman something in the get-up-and-go department. Bristow, however, was looking for something different. He had flown the Boeing 100, a commercial version of the P–12 fighter, for a short period and wanted something that could far out-

perform both the Boeing and the Stearman, yet he wanted to retain the trusty R–985 Pratt and Whitney.

Curtis sat down at his drawing board and designed what he thought was the smallest biplane that could logically be built behind the big Pratt and Whitney. Then Bristow came in with yet another list of little extras he wanted on the plane. These "little" extras included things such as a 50 gal. smoke tank (a Pitts Special uses a 5 gal. tank), 100 gal. of fuel to give it maximum range for cross-country (a Pitts Special has 15–22 gal. of fuel), and he wanted to be able to carry a steamer trunk on board so he wouldn't be constantly running out of clothes while on tour. Pitts Special pilots carry their toothbrush in their pocket and their extra pair of jeans in the turtle deck behind their head, and Jess Bristow wanted to carry a steamer trunk!

At first Curtis protested. In fact, he accused Bristow of having pushed negative g's one too many times and suffering brain damage. But he sat down at the drawing board again and redesigned the airplane to take in all of Bristow's seemingly impossible options. The various tanks were built into every nook and cranny of the fuselage, and the steamer trunk was fitted on board by removing the stick and snaking it up under the instrument panel where it was strapped to the back of the main fuel tank. The machine, when finished in 1951, was named fittingly enough, *Sampson*.

The Little Monster *used a C–85 Continental to generate speeds well over 230 mph.*

The brutish little bipe showed what could be done with a big round motor on a relatively small airframe. In its second incarnation, as built by replicator and air-show pilot Steve Wolfe, the airplane set several time-to-climb records.

Bristow campaigned *Sampson* for several years before it went through a series of owners and then met its bizarre death. Buddy Rogers was flying the air-

Little Monster, *as seen before a larger spinner was fitted circa 1951.*

plane, and on landing didn't notice an Ercoupe which had crawled into his blind spot. The Ercoupe and *Sampson* came together, with the 'coupe on the bottom, its prop cutting open the smoke tank and setting it on fire. Rogers managed to land the airplane, but the fire had too much of a head start and destroyed the airplane.

Sampson originally had a bump cowl, but Jess Bristow, the first owner, replaced it with a BT–13 cowl because the original took 150 screws to hold it in place. S. Wolf

A year or so had passed since the tragedy at San Diego and Curtis decided to get back into the racing game by moving the *Little Monster* to the front burner. Carrying race number eight, *Little Monster* was unique among many of the racers in that it had an all-metal wing and a three-longeron, steel-tube fuselage barely big enough to enable a small man to fit in it. In fact, Curtis says he had Quigley sit on the floor against the wall, and then drew around him to

determine the minimum "Quigley envelope."

The highly tapered wings used a built-up spar cap made of multilayers of aluminum strap riveted together with the layers all of different lengths. This was Curtis' way of dealing with decreasing bending moments without having to mill a tapered spar cap. As originally built, the hammered aluminum cowl came down to a small spinner, giving it a rather pinched-nose front appearance. About 1951, however, a larger 14 in. spinner was fitted and the cowling revamped to give it a much more pleasing line. All of the aluminum cowling and wheel pant work was done by Pitts, with the help of Phil Quigley and whoever

worked at or wandered into the shop. The pieces of aluminum were beat down into concrete forms and then welded together to form the completed components.

Curtis likes to tell of the time someone volunteered to hold the two wheel pant pieces together while he welded them. Being aluminum, the pieces got extremely hot and were burning the fingers of the volunteer. However, said volunteer was sampling heavily from a convenient bottle of Southern Comfort at exactly the rate necessary to numb his fingers. By the time the wheel pants were finished-welded, the volunteer's fingers were all blistered, but his face was all smiles.

Bristow flew Sampson *for less than a year before selling it to Ben Huntley.* S. Wolf

Although Curtis always performed the first test-flights and initial shakedown flights on all his airplanes, he gave the flying duties during the actual races to Phil Quigley, partially because Willie Mae had made it very clear how she felt about Curtis doing his own racing. Quigley raced the airplane in 1950 and 1951 before it became too difficult for Pitts to keep up with the racing and his expanding mechanic's school. He was also beginning to do a significant amount of business in crop dusting, so he couldn't continue racing.

The crop-dusting business came about through Jim Holland, a World War II pilot and nationally known aerobatic pilot. Curtis had modified some Stearmans for Holland's dusting service. Then, when Holland left for England, Pitts bought the business and instantly found himself active in dusting and spraying. It would stay that way for the next twenty years, and Holland would continue to be part of the Pitts Special legend.

Bill Brennan had been filling in whenever Quigley couldn't race *Little Monster*, so he struck a deal in which Curtis would supply the airplane, Brennan would do all the flying and transporting, and they would split whatever winnings were left over at the end of the year.

Against such wicked speedsters as Bill Falk's *Rivets* and the *Cosmic Wind Trio*, *Little Monster* didn't have much of a chance. It did, however, give them a run for their money and finished consistently in the top three, letting Brennan at least recoup all of his expenses.

The airplane was competed in most of the Goodyear races until they were finally shut down. Its last showing at an official race was in 1952 in New York. It was resurrected in the 1960s, and its last known trip around the pylons was in 1971 at Cape May, New Jersey, at the hands of Jim Dulin, who purchased the airplane in the early 1960s. *Little Monster* had been completely submerged during a flood prior to Dulin's purchasing the plane, so he completely disassembled and restored it prior to racing. A second restoration included blowing a new canopy bubble, after the airplane wound up on its back in a stiff cross wind. In the early 1980s Dulin was killed in a motorcycle accident, and four years later his widow sold the airplane to Jim Clevenger of Marion, North Carolina, who, as of this writing, has restored the airplane and still owns it.

Curtis Pitts built one other airplane during the 1950s that is all but unknown and was meant to be his personal cross-country high-speed machine. It was a two-place tandem, midwing airplane vaguely reminiscent of a large Formula 1 racer but with a six-cylinder, 250 hp 0-540 Lycoming in the nose. The wings were of wooden construction with a steel-tube fuselage, and Curtis sold the airplane before finishing it. Dubbed *Big Hickey* by the owner who finished it, the airplane surfaced several times at the Oshkosh, Wisconsin, air show before disappearing into a hangar outside Pittsburgh.

Almost all of Pitts' airplane-building activities of the early 1950s were confined to the first three Pitts Special biplanes, *Sampson*, two racers and *Big Hickey*. His dusting business was taking so much time at this point and the air-show business was getting so flat, he couldn't justify spending the time on any more Specials. As the decade drew on, however, that was to change.

Steve Wolf on takeoff in his faithful reproduction of Sampson. *He used the airplane to set several time-to-climb records. S. Wolf*

The plans on which all later Pitts are based, were drawn by Curtis while Pat Ledford built N8L. Here it is pictured at the 1962 EAA Rockford Fly-in. P. Ledford

Flatwing
Birth of a Homebuilt Legend

For the first decade after the war, the huge number of surplus airplanes made it unnecessary to think in terms of building your own fun machines. Through their air-show performances, *Little Stinker* and *Black Magic* had impressed a lot of pilots and whetted a lot of appetites, but nothing concrete would happen on the sport airplane front until pressure built to force Curtis Pitts into selling plans for the Pitts Special. Until that happened, the airplane continued to be a specialty machine justified only by air-show pilots.

Shortly after the war, the concept of building your own airplane had been completely smashed courtesy of a federal law which left the legality of homebuilt aircraft up to the local states. With few exceptions, the states fell in line and simply banned them. However, by 1948 that situation had been reversed through the efforts of George Bogardus and some Oregon homebuilders. That set the stage for the formation of the Experimental Aircraft Association (EAA) in 1953.

With the formation of the EAA, the build-an-airplane-in-your-basement movement had a central communications system going for it. Where there had been individual groups around the

country, there was now a way of communicating amongst them, which caused the entire movement to grow, which in turn increased the demand for homebuilt plans.

Curtis had given passing thoughts to offering the plans for sale, and had gone so far as to enter into an agreement with Jess Bristow to sell plans but nothing came of the agreement. For that reason, after Caro Bayley's *Black Magic* was finished, not another Pitts Special was built until the very end of the 1950s.

In the last half of the 1950s, Curtis decided that since he had so much dusting business to do in southern Florida and since the GI Bill business that supported his mechanic's school in Gainesville had dwindled, he would move his family and his business, setting up an operation on a small, personally owned strip just outside of Homestead, Florida. While working out of the corrugated hangars on the little strip he soon began to get an almost steady stream of requests by mail or in person for plans for the airplane. Several people, specifically Bill Dodd, an Illinois nursery man who wintered in Florida, were continually banging on the door of the hangar, asking Curtis to come out with plans for the plane so they could build one.

Even as pressure built for the plans, as late as 1958, the Pitts Special drawings were nothing more than a few sketches scattered around various corners of Curtis' duster shop. There were no for-

mal plans. Then in 1959 one of Curtis' duster pilots, Jim Meeks, gathered together as many of the plans as he could find and built a Special that had what was considered to be a monster motor in it. Essentially what he had was *Little Stinker* with a 170 hp Lycoming. He named the machine *Mr. Muscles*, and no airplane was ever more aptly named. With that tiny airframe and huge surplus of power, it was a pre-*Sputnik* spaceship.

Until Meeks built his airplane, the only Pitts that had been built from 1945 to 1959 were as follows:

Number One: 1945, crashed, not fatal

Number Two: *Little Stinker*, 1946, now in the Smithsonian

Number Three: *Black Magic*, destroyed, in-flight fire, nonfatal

Dean Case-built planes: N11J, N12J, 12JJ and two others, only one before 1960, whereabouts unknown

Number Four: *Mr. Muscles*, N39J, 1959, still flying but modified

With all the interest growing around the airplane, Curtis finally gave in and decided to do something about formulating plans. A friend of his, Pat Ledford, one of the ringleaders of the "get the Pitts plans done" movement, agreed to build a Special if Pitts would stop by and make changes, producing working drawings while Ledford was building the airplane.

Ledford finished the airplane in 1960. Powered with a 125 hp 0-290 Lycoming,

N8L was to have a tremendous effect upon sport aviation, aerobatics and Pitts' life in general. N8L was always just inside the hangar door at the Homestead strip, and Pitts made it available to any and everybody who wanted to fly it. Independent estimates from both Ledford and Dodd say somewhere upwards of 150-200 different pilots flew it during those early years which rapidly spread the fame of the airplane far and wide.

N8L was essentially a *Little Stinker* or *Black Magic* modernized and beefed up a little to take a somewhat larger engine. Technically called the S-1C (C for Continental), Curtis envisioned the airplane having a 100 hp 0-200 Continental in the nose which would make it the ultimate sport airplane. However, since N8L had the 125 hp engine in it, no one wanted to take a step backward even if the Special would be lighter with the smaller engine. Everyone wanted more horsepower. According to Bill Dodd, a number of his group who were preparing to build aircraft had already bought their 0-200 Continentals but never installed them simply because the horsepower race was on and they didn't want to be left behind. In fact, the horsepower race became so pervasive Curtis was hard pressed to keep the plans updated as people kept hanging larger and larger engines in it. Soon, what had started out as a 55 hp airplane wound up routinely having 180 hp stuck under that stubby hood.

The S-1C differed only slightly from *Little Stinker* in concept, but several dimensions changed. The wings got longer again, increasing 3 in. to 17 ft., 4 in., and the fuselage length grew by 3 in., mostly in the cockpit. The cockpit also was set at 19 in. inside the longerons, at the shoulder, which remained a standard dimension until the present time. Spars and a few fittings were increased in strength to take care of the speeds and weight generated by the new engine. The S-1C was to be the most popular homebuilt aerobatic airplane ever produced, as well as leading the pack in the sport biplane category with over 2,700 sets of plans sold and an estimated 500-800 airplanes finished.

Pat Ledford works on the 125 hp, 0-290 Lycoming engine in N8L. P. Ledford

Flying the S-1C (or any single-placed Pitts for that matter) is the closest any civilian is going to get to flying a bullet. In the first place, it's tiny when compared to anything remotely resembling a normal airplane. Its wingspan is only 17 ft., 4 in.; today there are radio-controlled model planes with larger spans than the S-1C's. But the wings seem to get even shorter when you step up on the left wing walk, swing one leg over the cockpit and slide down inside it. If you are wearing summerweight clothes and are in the 170 lb., 5 ft., 10 in. category, a stock S-1C will feel reasonably tight. If, however, it is wintertime or you are above average in weight or height, saddling up an S-1C is going to feel as if you have slipped down into a gopher hole it is so tight. And it's a gopher hole that is not particularly well furnished. Since weight is the enemy of all aerobatic airplanes, the only items to be found in the cockpit are those essential to flight.

Painted fuselage tubing and the stark white of the fabric is the full extent of the luxurious interior. Your feet disappear somewhere up ahead and can only be seen by ducking your head and peering into the dark hole under the instrument panel. The original S-1Cs were so short from the seatback to the instrument panel (29 in.) that many builders added 3 in. in length to the fuselage in the cockpit, something which Curtis did himself in plan modifications. This gave birth to the term long-fuselage Pitts, which is really a misnomer since they're all stubby.

Once you're ensconced in the cockpit, you'll find there are only four or five things over which the pilot actually has control. Stick and rudder, throttle, mag switches and carburetor heat are the only movable items at his or her disposal. Few Pitts have any more dials on the panel than is necessary to tell the pilot how fast he's going, how high he is and whether his engine is healthy or not. Again, if it doesn't need it, it doesn't have it. This includes starter and electric system. Although some airplanes were built with the ability to fire it up from the pilot's seat, that's the exception.

With the mag switched on, your feet pushing hard on the brake pedals and the stick stuck to your belly, yell "Hot and brakes" so the gentleman out front with the Armstrong (or is that strong arm?) starter knows the engine is ready to go before he swings the prop through. The second the engine catches there is no doubt in your mind there are an awful lot of ponies in an awfully small airplane, since every time a cylinder fires it is transmitted back to you in no uncertain terms. Pushing the throttle forward with the brakes locked actually causes one wing to dip slightly as the airplane rocks due to the torque. *This is one serious flying machine*, your brain says. And your brain is absolutely right.

It is during taxi out to the runway you are reminded instantly of the Pitts Special's reputation for being squirrely on the ground. Each and every movement you make with your foot is amplified to the point that simply staying in the center of the taxiway can be a chore. If, however, you're a sensitive pilot, you'll notice it isn't the airplane that is being squirrely. It's the driver. Because the airplane has so much control and reacts so instantly, your feet will feel as if they are made of cast iron, since you can't treat the rudders gently enough. After a few hours, you'll learn to gently stroke the controls, rather than actually moving them. If you ask the airplane to do something, it will do it almost before you realize you've asked the question. For that reason, when doing any ground-handling operations, the name of the game is to tiptoe—just barely tap the rudders and don't do anything precipitous like panicking and mashing a rudder to the floor. He who mashes a Pitts rudder winds up with a mashed Pitts.

When you reach the end of the runway you'll notice your heart is pounding a little, whether it's your first takeoff or your hundredth. On your first takeoff, your heart is pounding because you don't know what's going to happen next. On your hundredth takeoff, it's pounding because you do know what is going to happen next and you can hardly wait.

The most important thing to do at this point is to put your head back and look straight ahead. When you do, you'll notice that directly in front of you is nothing but the nose of the airplane, since it sits so steep. But while looking at the nose, try to notice what's in your peripheral vision on both sides. You'll see the edges of the runway form a triangle on both sides, with the point coming up against the side of the plane's nose. Your job from this point on, whenever taking off or landing, is to keep those triangles an equal size and shape. That will be your clue as to whether you're in the middle of the runway and whether you're headed straight or not. If you are not headed straight, your butt will be telling you, because you will feel it trying to slide one way or the other. It's the pilot's job to very lightly tap on those rudders which will keep the triangles even and his butt happy.

Now comes the fun part.

There is no other place in general aviation that a pilot will get the exhileration and sheer grin-power that comes out of a takeoff in a Pitts Special. Not even in the military does it exist, with the singular exception of a catapult shot off of a carrier. But of course, that's only possible with a lot of taxpayers' dollars.

Once you're out on the runway you're flat out of excuses, you can't taxi back—you've got to go for it. Looking straight ahead but concentrating on your peripheral vision, gently and very smoothly begin to straighten out your left arm, bringing the power up. But do not—repeat, do not—slam that throttle forward. The engine and propeller are awfully big for this little airplane, and impulsively hammering the throttle to the stop like you're in a 65 hp Champ will not only reward you with a 45 deg. left turn off the runway, but will probably increase your laundry bill at the same time.

Keep the stick sucked into your gut and keep the tailwheel down. You'll notice the instant the noise starts getting louder, runway lights begin moving fast. By the time the throttle is against the stop you'll be hypnotized by the fact that something is hitting you in the back and the world is flashing past at a dizzying rate. You should also be fixated on the triangles on either side of the nose, since you should be doing a little tapdance to keep the nose headed absolutely straight.

Realistically, absolutely straight doesn't exist. Not the first time out, anyway. Many first-time Pitts takeoffs are characterized by a slow-speed squiggle at the beginning, a couple of good-sized whoop-de-doos and one healthy swerve usually followed by another "Oh my gosh" swerve at which point the pilot sucks the stick into his lap and leaps off the runway whether he is ready to or not. The nice thing about a Pitts Special's takeoff on a wide runway is, almost

Curtis Pitts works on N8L. P. Ledford

regardless of how frantic and out of control you may get, it blasts off the pavement so fast that you'll be in the air before the panic sets in.

On a narrow runway, you don't have that luxury. Keep it straight or else.

The panic that generally sets in after a first takeoff is not actually takeoff panic. It's the awful realization you are going to have to come back and land. Ninety-nine times out of 100 that panic is totally unwarranted, but it's healthy panic nonetheless since there is nothing like being scared out of your drawers to keep you sharp throughout the rest of the flight.

The instant the airplane is off the ground you'll find the same thing that was said about the way it taxis is also

true about the way it flies: it mimics every move the pilot makes. If you wiggle your finger you will wiggle the airplane. This is an entirely new sensation to the average general aviation pilot who thinks in terms of displacing a control to get the airplane to do anything. In a Pitts, nothing moves in normal flight. Rather, there is a gentle pressure applied to each control to make it do what you want, since a visible movement of the stick of a half inch or more will slam the airplane into 90 deg. bank in a heartbeat.

Everything about a first flight in a Pitts is guaranteed to scramble the brains of a general aviation pilot. In the first place, even with a headset, the engine is pounding away with an exhaust tone reminiscent of a Mustang or an unlimited hydroplane. If the airplane has no canopy, the slipstream is ripping

at the back of your head reminding you it's there. The controls will feel so sensitive and the responses so immediate it gives rise to images of trying to balance a bongo board—one of those teeter-totter affairs where you stand on a short board trying to balance it on a roller underneath. If you can't picture that, try imagining hanging on to a wet watermelon seed.

The initial impression of a Pitts is just that—initial. Once the first-time pilot has had time for his nerves to settle down and his hands to stop shaking, he'll realize the airplane is connected directly to his brain and what he is having trouble dealing with is simply old-habit patterns: he is used to asking the airplane to do something and having to wait for it to do it. He is not used to the immediacy of control the Pitts gives him. Once he falls into the groove and relaxes his death grip on the stick, he'll begin to sample experiences unavailable elsewhere.

Although homebuilt aircraft are just that, homebuilt, and they vary in their flight characteristics, generalities about a flatwing Pitts can still be made. In the first place, once the pilot has leveled off, he has already begun to get an inkling about how to control the airplane. Or better yet, how not to. He has rested his right forearm on his right leg and barely fingers the controls. The airplane has very little natural stability. In fact, if the controls are released and the airplane is in perfect trim, it will still fall one way or the other in a matter of seconds. In 5-10 sec. the airplane is generally nose down, in a tightening spiral.

It is its lack of stability and willingness to answer the controls that make the airplane such a great aerobatic machine. The flatwing already wants to move in some direction, so, when the pilot suggests a direction, it is all too willing to comply.

Since the S-1Cs have only one set of ailerons in the lower panels, their roll rate is about 140 deg. per second. Put in perspective that is approximately twice as fast as a Cessna 152, but only two-thirds as fast as the later four-aileron Pitts. A rate of 140 deg. per second is, however, exactly the same as an S-2A two-holer. By any comparison, it is more than most civilian pilots have ever experienced.

It's not the roll rate that catches pilots by surprise, though. It is the break-out forces (or lack thereof) and roll acceleration that generally grabs them. Most airplanes require a slightly exaggerated pressure to get the controls out of the neutral position. This break-out force gives the feeling there is a notch at neutral that the airplane will generally fall back into. On the S-1C, that pressure is practically nonexistent. Consequently, something as minor as the weight of the pilot's hand becomes a disruptive force which moves the stick and asks the airplane to move.

Once the ailerons are deflected, most planes take just a second to develop a rate of roll, since they have to overcome the airplane's inertia. This is largely a function of wingspan and aileron effectiveness. In the case of the S-1C, it has very little span and lots of effectiveness. The result is that the airplane acceler-

ates to maximum roll rate like a lightning bolt. It is this unreal ability to rip from one back angle to another that boggles most new pilots' minds.

The ability to adjust one's attitude instantly is what endears the airplane to the akronuts of the world. Regardless of the airplane's horsepower, the S-1C still lets the pilot yank and bank until he is too pooped to want anymore. A simple aileron roll is a simple thought: *I want to roll, I think I'll move the stick.* An aileron roll happens faster than the words "I want to roll" can be thought.

The airplane's ability to climb is primarily a function of the horses available and the weight carried. S-1Cs are commonly seen with 150 or 180 hp Lycomings and less commonly seen with 125 hp engines. Although at least a few were built with 85 or 90 hp Continentals, none are still registered with that small engine. The urge for get-up-and-go caught up with all the original builders, and they re-engined their airplanes.

With even the 125 hp on board, the airplane has a climb rate far in excess of anything civilian that has propellers, but with the 150 hp and up, it is awe-inspiring with rates of 2,500 fpm (feet per minute) being common. In this respect, the flatwing airplanes with their more flat-bottomed M-6 airfoil have a performance edge over later airplanes with the same engine but the symmetrical airfoil.

The pilot is up having a wonderful time akroing his brains out, then the awful thought recurs, *I still have to land this thing.*

The landings are another part of the Pitts Special that have made the species something of a legend in sport airplane circles. The airplane is short, and the landing gear is narrow and stiff. Add to that the fact that short wings let the airplane come down like a shuffleboard puck, and you have an airplane that challenges a general aviation pilot's skills. Here, however, the same thing applies: The airplane goes exactly where you want it to go and does exactly what you want it to do. Nothing more, nothing less. If a pilot doesn't know what he wants to do with it, naturally the airplane is not going to know what it is supposed to do and it is liable to do something on its own that's stupid. It is up to the pilot to keep it from doing that. It goes back to the old adage, If you don't

want your airplane to go over there, don't let it go over there.

If the pilot is used to flying normal airplanes, he's probably also used to flying a rectangular pattern. This means flying a base leg 90 deg. to final, turning final, driving down final and then landing. In the case of the Pitts Special that means there will be a good-sized portion of final approach where he can't see the runway as well as he would like, although in the single-place airplane it is much easier than in the two-places. That's one reason so many Pitts drivers don't fly a rectangular pattern. Instead, they blend base and final together in a 180 deg. continuously turning approach that is almost identical to what WW II carrier pillots used to make to the carrier deck so they could see around the nose. In this way the pilot never loses sight of the end of the runway. It does, however, introduce another variable in that he has to vary his turn rate to put him on the centerline of the runway just as he runs out of altitude.

Many pilots have a tendency to fly the Pitts entirely too fast on final just to keep the nose down so they can see what they can. Speed of 85 to 90 mph is adequate for any of the models and, although it is not done by every pilot, it is a good idea to make as many of the approaches as possible with power off so that later on, should a problem occur, the pilot will know exactly where his airplane is going to wind up if the propeller is no longer propelling.

One of the problems introduced by a power-off approach, however, is the sink rate. The airplane is coming down at what appears to be a frightening rate. First time out, that can be a little disconcerting. Actually, it can be scary as hell. For that reason, the first few times out, it is not a bad idea to creep just a little bit of power (repeat, just a little bit) into it to make flairing easier. Otherwise, there is just too much happening at one time. Hold that power while flairing and, as you come into a three-point attitude, very gently creep the throttle back while lowering the plane to the pavement.

By initially using a little power in the approach, it gives plenty of time to make sure the airplane is pointed exactly straight. Repeat, exactly straight. Not sort of straight, not a little straight, but exactly straight, since a Pitts changes its personality completely depending on

The colors in this old photo of N8L may be fading, but the excitement the airplane helped start is still strong.

whether it touches down straight or crooked.

If all three gear touch at the same time while rolling perfectly straight, the airplane will continue to roll ahead more or less straight until it begins to decelerate down through 50 or 60 mph, at which point it will develop a mind of its own temporarily and swerve right or left with no warning. This is one reason most Pitts pilots will begin dragging the brakes almost as soon as they touch down in an effort to get the airplane to decelerate as quickly as possible.

If the airplane is allowed to touch down crooked, things get exciting immediately.

The most important thing to remember on touchdown is not to do anything jerky or hard. All movements involved in landing the airplane should be very gentle, with every movement limited in its scope, since any overt movement is going to reward the pilot with a sideways view of the runway. Again remember, the airplane will go exactly where your feet ask it to go—no place else. For that reason 90 percent of ground loops or trips into the bushes are the result of the pilot overcontrolling first one way and then the other in an attempt to correct the first swerve. If the nose swings off center in a swerve, don't try to bring it back to the center right off the bat. The first tap on the rudder should be to stop it in the direction it is headed. Then, tap it again but just enough to bring it back parallel to the centerline, although off center on the runway. If you try to bring it back to the centerline first, you'll build up a swerve that is going to make you feel as if you are doing a tap-dance in a mine field while trying to get straight again.

When landing a Pitts there is no substitute for anticipating the movement, correcting for it the second it is seen, and doing everything quickly but gently.

Incidentally, don't be concerned about little bounces or skips. The gear is so stiff and the wheels are so small, it is practically impossible to ease on the ground and stay there. Two out of three times, the airplane will skip on even the smoothest touchdown. Drop it a foot or bang it into the ground and it will reward the pilot with a world-class bounce that will require a touch of power to soften the return to earth. Without the power, the airplane will dribble down the runway in a string of bounces with decreasing amplitudes until it finally runs out of energy. Always an embarrassing way to arrive, but totally controllable and nothing to worry about. The pilot's primary job on touchdown is to keep it straight, since gravity always wins and the airplane will eventually stay on the runway.

One of the more entertaining things to do to a first-time Pitts pilot is ask if he or she can whistle after getting out of the airplane. Most of the time their mouths are so dry they won't be able to whistle for ten minutes. That will go away, as time is built up in the airplane, but it will come back from time to time, since a Pitts is always testing its owner to see how good he or she is.

Every Pitts pilot who ever lived can vividly recall that first takeoff and landing, and almost always rates it as one of the most exhilarating, worthwhile experiences of his flying career. Flying a Pitts for the first time gives new meaning to the phrase often seen on T-shirts at akro contests, "If it ain't a Pitts, it ain't sh—!"

S-1S Roundwing
Meeting the New Challenges

For several reasons, 1960 was a pivotal point in the history and development of the Pitts Special. The fact that plans for the S-1C were in the process of being made available to the akronuts of the world was important, but overseas something happened that had far longer reaching implications: The first World Aerobatic Contest was held in Budapest, Hungary.

Until the Budapest competition, aerobatic competition had been largely a national sport confined to each nation and loosely conforming to the rules that nation applied. With the recognition of aerobatics as a world-class sport, however, an international committee was formed and a set of rules and definitions of maneuvers developed that would serve as guidelines for all competing nations.

Paralleling gymnastic scoring systems, each maneuver was given a difficulty factor and a grading scale; then

The S-1S roundwing of Bill Thomas was a familiar sight during the world contests of the late 1960s and early 1970s. It typified the Pitts that beat the best the world had to offer.

Pit Row at the World Contest in Hulavington, England, 1970. The Roundwing Pitts of Bob Herendeen, Bob Schnuerle and Gene Soucy stand ready to take on the world.

Known, Unknown and Compulsory sequences were developed which each competitor had to fly. This gave every nation the same basic set of rules to fly by.

The original sequence development and scoring system was developed by Spain's Count Aresti, hence the name Aresti System for the aerobatic shorthand used in choreographing maneuver sequences. Aresti flew a Jungmiester, which because of its high drag and low 160 hp was especially weak in doing vertical maneuvers. So, he graded vertical maneuvers higher than others. Also, since his Jungmiester had flat-bottomed wings and couldn't do outside or negative-g maneuvers very well, they were also graded high. The result is that competition placed emphasis on doing vertical and outside maneuvers well, if a maximum score was to be realized.

Frank Price, a well-known air-show pilot from Texas, decided that, since America didn't have a team and there was no organization to put a team together, he'd become the country's akro representative. Crating his aging 185 hp Warner-powered Great Lakes biplane, he traveled to Budapest to carry America's aerobatic flag.

Besides being woefully underfinanced (he had to leave his airplane in Europe after the contest until he could raise money to get it back), his Great Lakes wasn't a match for the new generation of European airplanes. It was too big

and didn't have the power-to-weight ratio needed to work the highly graded vertical maneuvers. Still he tried. As if fitting into the Texas legend, he felt something had to be done, so he did it. In pioneering fashion, he fought the fight with what he had.

The most important factor of the Budapest competition to future akro efforts was that it gave Price a fountain of aerobatic competition information to bring back to America and help in formulating future aerobatic policy.

Almost as soon as the Budapest competition was announced, Curtis Pitts perked up his ears and started sponging in information. With Price's return, his suspicions were confirmed: The S-1C would be an excellent competition airplane, but would need modifications to work the outside maneuvers and needed more roll rate for the verticals.

The M-6 airfoil on the S-1C was a good choice at the time the airplane was designed. It was a favorite among many designers, but it was fairly flat-bottomed, which meant it wanted to lift more in one way than the other. In inverted flight, the plane needed more angle of attack to maintain flight than it did right side up, which means it generated more drag and generally wasn't as good upside down as it was right side up. It was still much better than anything else in America, but the writing was on the wall: The Europeans were going to develop specialty aerobatic airplanes,

With nothing to judge scale, the S–1S appears to be normal size. It doesn't shrink until an average size pilot stands next to it.

and their emphasis would be on outside and vertical maneuvers.

Curtis' design goal was simple: He had to modify the S-1C until it didn't know, or care, whether it was right side up or upside down.

The M-6 airfoil had to go. But, it wasn't simply a matter of grabbing a symmetrical airfoil and making wings that fit. He found that out almost immediately.

Shortly after 1960, using Pat Ledford's airplane, N8L, as a test bed, Curtis began designing and building symmetrical wings for the airplane. Using the first airfoil selected, the airplane flew just fine and was greatly improved upside down—that is, until Curtis put hard negative g on it! Then, to use his own words, "it just went sort of crazy and I nearly went in...." After that flight, Pitts grounded the airplane and wouldn't let Ledford fly it again. He began taking the wings off that afternoon.

After several other tries, he came up with a system of using a different airfoil for the top wing than the bottom, which among other things let the airplane stall the same when upside down that it did right side up. The final selection of airfoils, which were actually developed by Curtis himself, and the way in which they were used formed the basis for a patent.

The final set of wings let the little Pitts Special ignore which direction the pilot's head was pointing. It didn't know up from down and could perform any maneuver from any position, regardless of whether it was inside or outside.

At this point, Curtis decided he needed to look at the complete airplane and make sure it fit into the competition envelope in every possible way. Where it was structurally weak, it was beefed up. The fuselage was lengthened a total of 12 in. over the original plane, which included three welcomed inches in the cockpit. The additional length made the airplane groove better in maneuvers and also made it somewhat easier to

52

land. The longer fuselage also solved the center of gravity problems associated with hanging an increasingly heavier and more powerful series of engines in the nose.

The upper wings were modified to receive a set of ailerons, which were operated by slave struts from the lower ones. Initially, these were normal-appearing Friese ailerons, which in later years gave way to the so-called symmetrical ailerons that were actually thicker than the wing, but boosted roll rate because of their better aerodynamics.

The fuselage was beefed up to accept the 0-360, 180 hp Lycoming as the basic engine. With this engine and a loaded weight of approximately 950 lb. (715 lb. empty), the airplane was carrying a shade over 5 lb. per horsepower. Only a lightly loaded F8F Bearcat could boast of a lower power loading. This put the airplane into the missile category, when it comes to leaving the ground behind.

The new airframe-wing combination was officially designated the S-1S, for symmetrical. However, the entire world rapidly began to call it the roundwing Pitts, because of the distinctive belly to the wings.

The symmetrical wings were designed specifically for the competition pilot. In fact, they were never made available in plan form until well into the 1970s, when S-1S plans were made available by the factory. But that's a story for a later chapter.

By 1964 Curtis Pitts was so busy with aerobatics that he sold his duster business to concentrate entirely on the little airplanes. When he made that decision, he also made a decision to operate like a business, which meant marketing his strongest products exclusively. This was one reason the roundwing wings were never available as plans; he sold them only as completed units or on complete airplanes he had built. At the time, the

The Lexan windows on the side and belly of an S–1S allow the pilot to better orient him or herself in all attitudes.

wings cost $2,500 and a complete roundwing 180 ran $12,500 which sounded like a lot of money, but not if a pilot was serious about winning at aerobatics.

Many of the S model modifications were carried over to the plans for the homebuilder; the only thing missing was the symmetrical airfoil. The four ailerons and longer, stronger fuselage were incorporated into a new model known as the S-1D, a designation seldom used, since most folks still refer to that model as a long-fuselage, four-aileron flatwing.

One of the first roundwing Pitts went to US national champion Bob Herendeen. Herendeen had been competing in a flatwing, including the world contest in Moscow in 1966. When he heard a symmetrical wing was available, he

Factory built S–1Ss will be found in all paint schemes and with or without electrical systems.

immediately ordered a new airplane, which carried the number N266Y. His flatwing had been N66Y. An airline pilot by trade, Herendeen was one of the first serious competition pilots who looked at the world championships as something to be conquered. Herendeen and his plane were to become the most important factors to the Pitts Special's growth in popularity in the post-*Little Stinker* era. Taking delivery of N266Y in February of 1968, Herendeen easily captured the national championship in 1969.

In 1970 at the world contest in Hullavington, England, Herendeen was forecast as an easy winner, when his engine quit for no explained reason in a spin. He had to zero the rest of the flight.

The cockpit of an S–1S is snug and crowded. The Aresti card on the panel choreographs the pilot's moves.

When he came back to re-fly it, they didn't give him credit for the spin from the first flight and the difference cost

him the championship. He came in second behind a Russian.

Although luck wasn't on Herendeen's side and he was never to win the world crown, he did establish one absolute fact: the roundwing Pitts was the airplane to beat. Politics had always played a large part in the eventual winners of the contests, since it always seemed the host country's pilots were favored. When Herendeen was through with his world sequences, however, the performance difference between the Pitts and the rest of the airplanes was so large, there was no doubt the next time around even politics couldn't close the gap.

Where Herendeen's airplane was the first competition roundwing, it was only a matter of months before the little Homestead, Florida, hangar cum airplane factory was humming, hammering out wings for everyone who was serious about winning. Three of the early recipients of wings or complete airplanes included Charlie Hillard (his was built by Bob Heuer), Gene Soucy and a young guy named Tom Poberezny. In a few years, these three became the Red Devils and then, when Frank Christensen entered the aerobatic picture, they became the Eagles. It's interesting to note that between these three pilots, they hold five national championships (Hillard one, Soucy three, Poberezny one) and one world championship (Hillard).

One of the early airplanes also went to one of Curtis' neighbors, Mary Gaffaney. Gaffaney, along with Charlie Hillard, went on to write a few aerobatic championship chapters of their own.

The US World Aerobatic Team for 1971 included Hillard, Soucy, Poberezny, Bill Thomas and Art Scholl on the men's side, while the women's team included Gaffaney, then reigning US Champ, and Carol Salisbury. With the exception of Scholl and Salisbury who shared the same S-2A, all were flying S-1S Pitts

This factory-built S-1S shows the longer fuselage and symmetrical airfoils that identify the roundwing models.

Next page
Not all S-1Ss originally had canopies. Charlie Hillard's world contest version was the first Pitts to fly competition with an enclosed cockpit.

National Champion Bob Herendeen with first world competition roundwing. This same airplane was used by Betty Stewart to become the only person to win two consecutive world championships. EAA

Specials. Hillard's plane was the first single place with a 200 hp Lycoming swinging a fixed-pitch prop. His was also the first to use a canopy during competition flight.

The 1972 world contest was held in Salon de Provence, France, and was to prove pivotal in many ways. The rivalry between the Russians and Americans had become intense. In fact, aerobatics in general had become an East versus

Gene Soucy (left) and Tom Poberezny (right) and their roundwings helped win the 1972 Men's World Aerobatic Championship. EAA

In addition to helping win the team championship, Charlie Hillard won the men's individual championship. EAA

Both sides had been working feverishly to build better airplanes and train better pilots. Salon de Provence was the rough equivalent of the dusty street in some lonely western town, with the good guy and the bad guy standing face to face, hands poised.

The flying was fierce and accurate and the judging, which has always been hotly contested by whichever side was on the losing end, was thought to be as fair as had been seen to date. This could have been the result of the contest being on more or less neutral ground.

When the smoke cleared, the Texan with the red Pitts was still standing, and alongside him was the woman from Florida, backed by the entire US team—it was a clean sweep. Americans had won everything: both men and women's titles (Hillard and Gaffaney) and the team title. The Pitts Special reigned supreme.

By this time there were hundreds of Pitts Specials running around the country, the vast majority of which were the standard, flatwing S-1C. The round-wings were predominately the airplane of the competition pilot and the flatwing slowly became recognized as one hell of an airplane, but since competition had passed it by, it was relegated to the competitor who was only interested in the lower levels, through intermediate category. Although the flatwing was still competitive in the advanced category, it was obvious the top dogs were spending the money and re-equipping their airplanes with roundwings, not all of which came from Curtis Pitts.

A number of roundwing clones were suddenly offered to the public in kit form which made it possible for the homebuilder to get the latest in Pitts technology. Because of their great similarity to the factory wings, Curtis is convinced a set of drawings disappeared from his shop in someone's lunchpail and he thinks he knows who, but can't prove it. At any rate, a number of imitators sprang up, the most successful of which was that offered by Sparcraft. The primary difference in the Sparcraft wings was that the ribs were plywood rather than built-up spruce trusses. It's unknown exactly how many Sparcraft roundwing kits were sold, but it has to number in the hundreds.

The 1970s were the golden era for the Pitts Special. Pilots such as Betty Stew-

West battle, with the heavily state-subsidized teams of the Soviet Bloc having obvious advantages over the western teams, almost all of which depended upon donations and private money. The Czechoslovakian Zlin 526s and Russian Yak 18s, both monoplanes, were designed for a specially selected cadre of pilots who had nothing to do but train to become the best aerobatic pilots in the world, and in so doing, beat those from the West. Especially the Americans.

art racked up record after record, she being the only pilot, man or woman, to ever win world championships back to back. The Red Devils, who did their first impromptu show at the world contest in France and their first professional show at Transpo 1972 in Washington state, flew nearly 350 shows over the next seven years giving the Pitts tremendous visibility before trading in their beloved Pitts for their sponsor-built Christen Eagles.

During this period, the Pitts reigned supreme, but a change was written on the wind. A young Leo Loudenslager had shown what he and his highly modified Stephens Akro monoplane could do at his first national contest early in the decade. By the end of the decade, his Laser 200 had become the airplane to beat, although every slot below him in contests was filled by a Pitts. The Pitts was still numerically king, but the very pinnacle was slipping away from it.

The US women's team was also victorious at the 1972 world championship, and Mary Gaffaney flew her way to the women's individual championship. It was a clean sweep for the US teams and for the Pitts Special. EAA

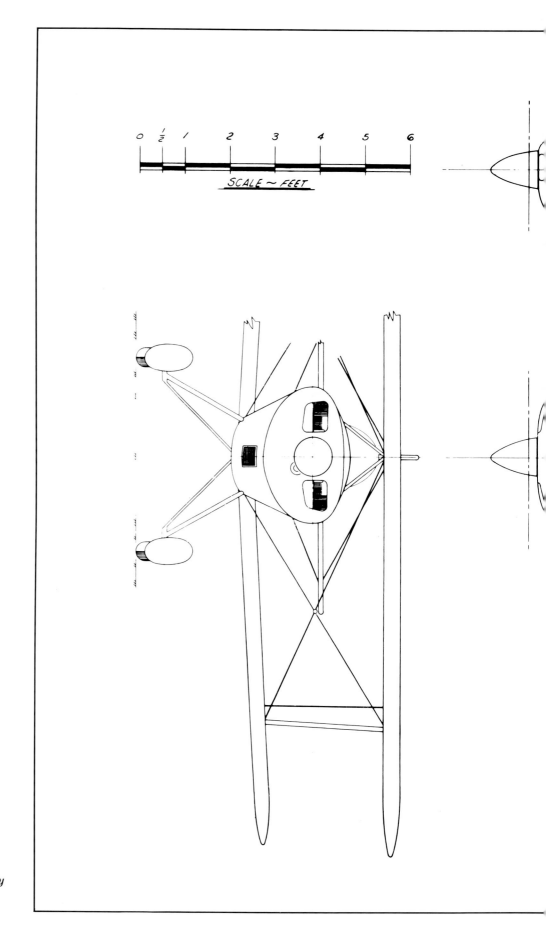

SCALE ~ FEET

Although it is the final major airframe variation of the S–1, the S–1S could clearly trace its roots back to the first Pitts in 1945. Christen Industries

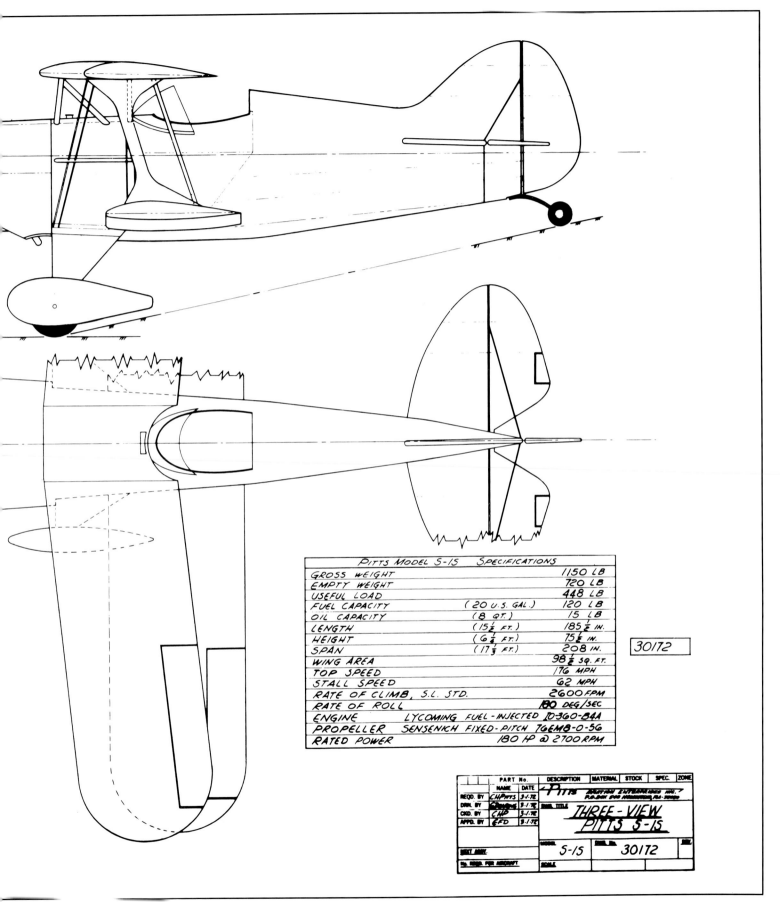

PITTS MODEL S-1S SPECIFICATIONS		
GROSS WEIGHT		1150 LB
EMPTY WEIGHT		720 LB
USEFUL LOAD		448 LB
FUEL CAPACITY	(20 U.S. GAL.)	120 LB
OIL CAPACITY	(8 QT.)	15 LB
LENGTH	(15½ FT.)	185½ IN.
HEIGHT	(6¼ FT.)	75¼ IN.
SPAN	(17⅓ FT.)	208 IN.
WING AREA		98½ SQ. FT.
TOP SPEED		176 MPH
STALL SPEED		62 MPH
RATE OF CLIMB, S.L. STD.		2600 FPM
RATE OF ROLL		180 DEG/SEC
ENGINE	LYCOMING FUEL-INJECTED IO-360-B4A	
PROPELLER	SENSENICH FIXED-PITCH 76EM8-0-56	
RATED POWER	180 HP @ 2700 RPM	

30172

	PART No.	DESCRIPTION	MATERIAL	STOCK	SPEC.	ZONE
	NAME	DATE	PITTS	AVIATION ENTERPRISES INC. P.O. BOX 548 HOMESTEAD, FLA. 33030		
REQD. BY	CHPITTS	3-1-72				
DRN. BY		3-1-72	DWG. TITLE			
CKD. BY	CHP	3-1-72	THREE-VIEW PITTS S-1S			
APPD. BY	CFD	3-1-72				
NEXT ASSY.			MODEL	DWG. No.		REV.
TO. REQD. PER AIRCRAFT		SCALE	S-1S	30172		

S-2

Birth of the Two-Hole Pitts

How does a pilot learn what it looks like to be going straight up, while doing a snap roll? Or how does he or she master the intricacies of doing inside-outside rolling 360s? Until the early 1960s, when aerobatic competition heated up and the scoring methods became so precise, the accepted method was to talk about it on the ground; then the pilot went up and kept trying until it looked right. That wasted a lot of time, fuel and elbow grease.

As the little Pitts Specials began pouring out of workshops, both Curtis' and others', and began cavorting around in high-g maneuvers, Curtis Pitts recognized the problem of training advanced aerobatic pilots. The basic maneuvers could be learned in practically anything, but the serious stuff took a serious trainer, something that didn't exist in America. There was a huge gulf between anything capable of carrying both a student and an instructor and the single-

The S-2 made it possible for a much wider variety of pilots to sample unlimited aerobatic flight and did much to popularize the sport.

The master himself in the prototype S-2 at the last EAA fly-in held at Rockford, Illinois, in 1969. B. Frankmans

place airplanes being used in competition. In fact, there wasn't a single two-place airplane available that was capable of doing even a vertical roll, much less any of the outside maneuvers.

Even as he was developing the round-wing single-place, Curtis already had a two-place airplane bubbling away on his mental back burner. By 1965-66, it was moved up to the front burner and became project number one.

The two-place airplane was going to be an entirely different type of venture. By far, the most ambitious part of the venture was that the airplane would not only be capable of doing all the unlimited category aerobatic maneuvers, but it would also carry FAA pedigree papers declaring it Type Certified and therefore capable of being used as a commercial airplane for instruction. That one feature, the Approved Type Certificate (ATC), would allow flight schools to use the airplane, thereby opening advanced aerobatics to a much wider audience. On the other side of the coin, going for an ATC meant Curtis had to jump through all the FAA regulatory hoops that apply to any factory-produced plane. The single-place airplanes were all in Experimental category, so none of the normal FAA rules applied. In short, he was about to jump into an arena where even the established factories feared to tread. Here he was a duster operator on a small grass strip in Homestead, Florida, preparing to be the first person since the 1930s to design and

certify an aerobatic biplane! The word ambitious is certainly apropos.

Curtis is nothing, if not a bulldog. When he gets his teeth into something, he doesn't turn it loose until it either works the way he wants, or he has proven it isn't feasible. He knew the two-place, known simply as the S-2, was feasible and he wasn't about to turn it loose.

The S-2 could not be simply a stretched version of the S-1. The extra 3 ft. needed to house another person, and all the controls meant designing an entirely new plane. The new airplane would have a span of 20 ft. and length of 17 ft., 9 in. This represents a sizeable increase over the single-place planes. Although at 20 ft. it still looked like a model airplane to most pilots, it would look like an airliner compared to the single-hole birds.

Since the airplane was to be an unlimited aerobatic airplane, it had to be able to climb the Aresti ladder of maneuvers with no limitations whatsoever. It had to be perfectly at home in any attitude. That meant the airplane had to have the same stall and slow speed characteristics upside down that it had right side up. Traditionally, this had been a nearly impossible goal for the biplane designer.

Normally a biplane was designed so the top wing had more angle of incidence, which caused it to stall first. Since the lower wing would still be lifting after the top one had stalled and was behind the center of gravity, it always

*A moment in Pitts history. Right to left is
Gene Deering, engineering representative
on S–2, Curtis Pitts, Tom Poberezny, Bob
Schnuerle in back cockpit, Bob Herendeen
and the author in front hole of prototype
S–2. The year was 1971.*

caused the nose to drop in a stall.
Inverted, however, that same arrangement can cause the opposite to happen—the nose doesn't want to come
down until both wings have totally
stalled.

Complicating the matter is the fact
that unlimited aerobatics automatically
demands a symmetrical airfoil so the
wing lifts the same in either direction.
No biplane had ever been designed with
symmetrical airfoils on both wings, prior
to Curtis' experiments on Pat Ledford's
single-place. In the course of his experiments, he found he could use different,
specially designed airfoils top and bottom, and control his stall that way and
not compromise the airplane's aerobatic
capabilities. This concept would be the
heart of the S–2's design.

Obviously, when designing an airplane
to carry two people and maximize performance on a given engine, the airplane
itself has to be minimized. It has to be as
small as possible. However, the FAA
standard pilot, 5 ft., 10 in. and 170 lb.,
can only be compressed so much. Pitts
snuggled the pilots together as much as
possible, putting the front seat between
the pilot's legs, so the rudders for the
back seat were actually in line with the
passenger or student's hips. Everything
was an exercise in compactness, which
kept the airplane as small as possible.

When it came to stuffing horses ahead
of the firewall, Curtis again had to keep
it light, while maximizing the horsepower. A number of engines were looked
at, including the 220 hp Franklin 0–360.
It was powerful and smooth, but it was
also six cylinder, which made it too long
to work with the weight and balance of
the airplane.

*Curtis Pitts and Betty Skelton in N22Q at
Rockford. FAA*

The engine selected was the tried and true 0–360, 180 hp Lycoming, which was showing up in the single-place Pitts with regularity. The engine was modified for inverted flight with the addition of inverted fuel and oil systems.

When the airplane rolled out of the shop it sported a red, black and white sunburst paint job and the image of a pudgy little skunk on its turtle deck. It had been dubbed *Big Stinker*.

The airplane was test-flown for the first time in mid-1966, just in time to make it to the EAA Convention at Rockford, Illinois. Curtis made the first flight and remembers, "I actually overshot the first landing and had to go around. It was a whole lot slipperier than I expected it to be. It just didn't want to slow down on final."

With an empty weight of 887 lb., it was nearly 150 lb. heavier than its single-place siblings. With the extra weight and size, naturally it couldn't roll or climb with the S–1S, but it could still do all the maneuvers with ease, which meant there was now a two-place airplane that would give an aspiring akronaut a front-row seat on unlimited aerobatic training.

Getting the airplane flying was one thing, getting it certified was an entirely different matter. Early in the project, Curtis hired Gene Deering to be his Designated Engineering Representative (DER). He was to do the official engineering work and run interference with the FAA. By coincidence, Deering had also been one of the many Florida engineering students who had toured Curtis' mechanic-training facility in Gainesville nearly fifteen years earlier.

To get an accurate picture of what certifying the airplane entailed, it is necessary to first put it in context. The year is 1967 and the FAA is used to doing engineering certifications on the new series of jet airliners. The push is toward solving problems with increasingly sophisticated engineering and fabrication techniques. Performance is measured in Mach and distances in continents. Then they get an application for a rag-and-tube biplane that carries 22 gal. of fuel and is designed specifically to fly upside down.

The stories that came out of the certification and flight-test procedures are legendary. The level of the Pitts team's frustration with bureaucracy can be judged by the fact that after a long series

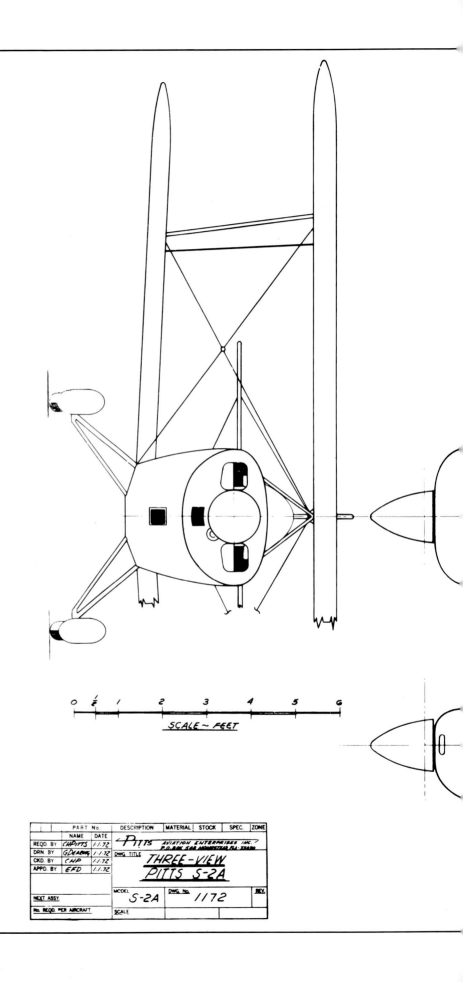

SCALE ~ FEET

		PART No.	DESCRIPTION	MATERIAL	STOCK	SPEC.	ZONE
	NAME	DATE					
REQD. BY	C.H.Pitts	1.1.72					
DRN. BY	G.Deering	1.1.72					
CKD. BY	CHP	1.1.72					
APPD. BY	EFD	1.1.72					

Pitts AVIATION ENTERPRISES INC.
P.O. BOX 548 HOMESTEAD FLA. 33030

DWG TITLE
THREE-VIEW
PITTS S-2A

MODEL S-2A   DWG. No. 1172   REV.

NEXT ASSY.

No. REQD. PER AIRCRAFT   SCALE

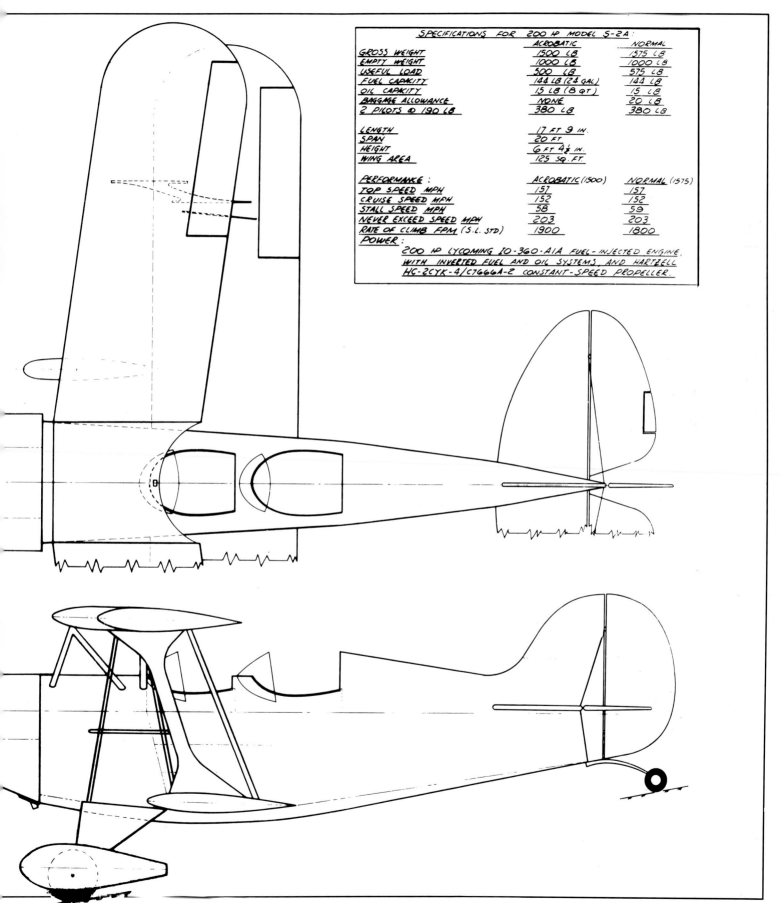

SPECIFICATIONS FOR 200 HP MODEL S-2A:

	ACROBATIC	NORMAL
GROSS WEIGHT	1500 LB	1575 LB
EMPTY WEIGHT	1000 LB	1000 LB
USEFUL LOAD	500 LB	575 LB
FUEL CAPACITY	144 LB (24 GAL)	144 LB
OIL CAPACITY	15 LB (8 QT)	15 LB
BAGGAGE ALLOWANCE	NONE	20 LB
2 PILOTS @ 190 LB	380 LB	380 LB

LENGTH	17 FT 9 IN.
SPAN	20 FT
HEIGHT	6 FT 4⅜ IN.
WING AREA	125 SQ. FT.

PERFORMANCE:	ACROBATIC (1500)	NORMAL (1575)
TOP SPEED MPH	157	157
CRUISE SPEED MPH	152	152
STALL SPEED MPH	58	59
NEVER EXCEED SPEED MPH	203	203
RATE OF CLIMB FPM (S.L. STD)	1900	1800

POWER:

200 HP LYCOMING IO-360-A1A FUEL-INJECTED ENGINE, WITH INVERTED FUEL AND OIL SYSTEMS, AND HARTZELL HC-2CYK-4/C7666A-2 CONSTANT-SPEED PROPELLER.

The extra cockpit in the prototype S–2 demanded the airplane be much larger, so the name Big Stinker seemed to fit. The airplane is now in the EAA Museum.

N22Q, the prototype S–2, used a 180 hp Lycoming with a fixed pitch prop and weighed less than 900 lb.

The S–2A was originally open cockpit, but almost all later airplanes were equipped with either the two-place or single-place canopy options.

of paperwork delays, the primary engineering document given to the FAA for drop-testing the airplane's landing gear contains a little humor that has probably gone undetected to this day. If a piece of paper is laid over the front page leaving the first letter of each line of copy showing, reading the exposed letters vertically advocates doing something sexually impossible to the feds.

The FAA assigned Joe Thompson as their official evaluation test pilot and Bob Schnuerle, a member of the US Aerobatic Team, acted as the Pitts factory pilot. Curtis remembers Thompson as a "really nice guy that learned to handle the airplane fairly quickly...." This is amazing, considering Thompson was a former Air Force lieutenant colonel and had been responsible for the acceptance testing on the B–58 Hustler. Quite a change in assignments!

Curtis openly thanked the FAA Engineering and Manufacturing District

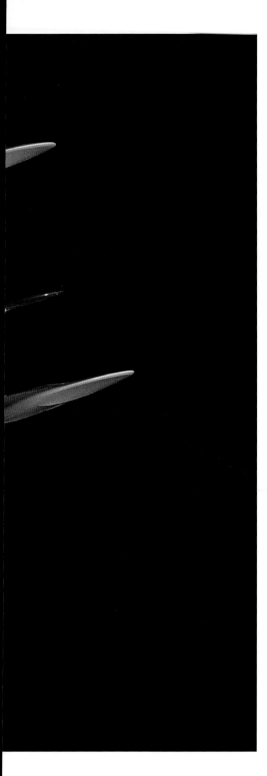

Office chief, John Vogel, for his help in keeping his subordinates in line. Often, when the field personnel would make a request or turn down something that was clearly unnecessary, Vogel would step in and let a little common sense prevail.

From 1960 on, when activity on the little airplanes had begun to pick up again, the entire Pitts Special saga was taking place in the most unimpressive-looking corrugated metal hangars imaginable. The Homestead, Florida, location was an 1,800 ft. long grass strip that bordered an orange grove and challenged a pilot's ability to even find the place, much less land on the 1,200 ft. that was actually usable. The hangar had been adequate for keeping a dusting business going, but, by the time even two or three of the tiny single-place airplanes were in the building process, the place was jammed.

With the two-place airplane destined to become a certified plane offered to a much wider market, Curtis was going to have to do something about production space. From the very beginning of the two-place project, he had planned on certifying it and then either selling the

project or joint venturing the production with another party or company. He had never envisioned his corner of the orange grove as being the proper location for a full-fledged airplane factory.

Unknown to Curtis, however, in Afton, Wyoming, events were taking place that would have a direct impact on his airplane and his future.

Afton was the home of the Callair series of agricultural aircraft and the plant was being run by Herb Anderson. Anderson was an old hand at building rag-and-tube airplanes, going back to

The factory-built S–2As all used a fuel injected, 200 hp Lycoming with a constant speed prop which is designated by the A in S–2A.

his days in Sacramento, California, where he went directly from high school to working with a company rebuilding Travelairs, WACOs and so on for dusting. When Stearmans became available after the war, he was deeply involved in converting many to dusters and sprayers and eventually in an ongoing business rebuilding wings for other operators. After a short stint at Callair in Afton, Anderson went on to become the production supervisor and eventually manager for Mooney Aircraft, in Kerrville, Texas, just when they were bringing out their soon-to-be successful series of four-place aircraft.

Anderson had a natural head for making production lines run. He knew how to fit the many parts of a manufacturing facility together so the small parts came in one door and a finished

The single rear canopy was originally certified by Art Scholl Aviation and was so popular the factory later added their version to the option list. The airplane could be flown as single place with front pit covered.

product flowed out the back. It's a rare talent. Recognizing this, Pug Piper (yes, of that Piper family) hired Anderson to be the assistant plant manager for the new Vero Beach, Florida, Piper plant and again, this put him on the team that brought an entirely new series of aircraft to the market—the Piper Cherokee.

The cool mountains and green valleys of Afton, Wyoming, were still in the back of Anderson's mind, however, and the Florida humidity quickly became more

than he could take. Fortunately, shortly after he had left Piper to set up his own business in Salt Lake City, the new management of Callair called on him to help bring them out of receivership, which he did. From 1962 to 1966, Anderson and his people at Callair built nearly 800 Callair A-9 agplanes!

When Rockwell-Aero Commander bought out the Callair line, Anderson once again found himself in the humidity of the South, this time in Albany, Georgia, as plant manager of that location for Aero Commander. It was on a trip to Homestead to consult with architects building a new Aero Commander facility that the Wyoming-Pitts Connection was initiated.

Always a biplane lover, Anderson decided to look up Curtis Pitts at his little strip at Homestead. He had read about the Pitts Specials and hoped to meet Curtis just to make his acquaintance. However, when Anderson arrived, no one was at the airport and the two-place Pitts prototype was sitting outside the hangar.

Anderson recalls, "I rounded the hangar and my heart started skipping a beat the second I saw the two-place Pitts."

To say Anderson liked the airplane is an understatement. By 1967, when the two-place had been flying for some time, he and Curtis Pitts had several conversations about the airplane, none of which were conclusive. Pitts was still heavily involved in getting the airplane certified and Anderson had been called back to Afton by Doyle Child, a local businessman who had bought the now-vacated Callair plant.

The Callair buildings were leased to the Polaris snowmobile company and Anderson was made general manager. It was in that environment that he was able to put his concepts about high-quantity, low-cost production planning into operation. However, when the highly successful company was bought by Textron, the decision was made to incorporate the Afton operation into one of their existing facilities in another state. This time Anderson stayed in his

mountains and formed a repair company he dubbed AeroTek.

With his plant buildings again empty, Child cornered Anderson and asked him for his ideas about utilizing the plant. The workers who had all those years of experience in building rag and tube airplanes still lived in the area, and the buildings were still usable. The question was what to use them for.

Manx Kelly of England was one of the first to capitalize on the S–2A's air-show capabilities and formed his Carling team around them.

Anderson thought about Pitts and his little airplane, which had by that time been certified. He got on the horn to Homestead and received a typical reac-

Previous page
The Carling team of the early 1970s doing
what it did best.

tion from Curtis, who said (slowly), "I'm going up to the Nationals in my Bonanza and I'll come by and see what you've got."

Doyle Child, Anderson and Curtis sat down for breakfast and got right to the bottom line: Child said he'd put up some money and give Curtis a rental rate on the buildings he couldn't afford to pass up. Curtis said he'd put in a similar amount, and the deal was done.

The venture was structured in such a way that Anderson's company, AeroTek, would do the manufacturing of the airplanes in Afton and Pitts Aviation Enterprises, based in Homestead, would do all the engineering and marketing.

Curtis spent several months sending drawings, jigs and templates to Anderson in Wyoming. Then he spent an additional several months in Afton physically overseeing and helping Anderson get things rolling. In the summer of 1971 the first S–2A rolled off the production line, with the initial airplane going to well-known aerobatic and air-show pilot, Marion Cole. Number Two went to

Art Scholl, himself a legend in his own time.

Anderson and Pitts accomplished a minor miracle: They had certified and put into production a fully aerobatic airplane. They had overcome obstacles that would have cost corporations millions of dollars, but they steamrollered each obstacle with experience, innovation and pure dogged determination.

The airplane that came off the production line was slightly different from the prototype, N22Q. Most of the differences were ahead of the firewall where the 180 hp Lycoming had been replaced with a fuel-injected, 200 hp, IO–360 Lycoming swinging a Hartzel constant-speed propeller and equipped with a full inverted fuel and oil system. Although the bigger engine gave the airplane an extra twenty horses and the obvious advantages of a constant-speed propeller, it also increased the weight from the 887 lb. of the prototype to 1,050 lb. The A in S–2A denotes these changes from the prototype S–2

The Pitts S–2A not only became the standard for advanced aerobatic training, it also set the stage for the factory to establish a long line of aerobatic airplanes that are in a class, category and realm all to themselves.

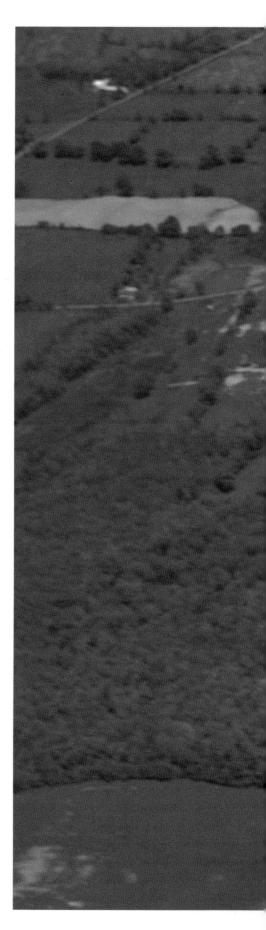

16 Papa Sugar
A Personal Essay

Every author gets to indulge himself now and then—and it's my turn now. It is absolutely necessary I step out from behind the word processor, because I simply can't speak of the Pitts S-2A in the third person. When it comes to that little 200 hp biplane, I have to speak in terms of "me" or "I", if only because one has occupied a major portion of my life for over twenty years. So it is understandably much more than a machine to be dissected or a flight to be described in abstract terms.

Pitts Special 16 Papa Sugar (the FAA insists it's Papa-Sierra, but we won't give in) came to live with me and my partners in September of 1971. It was the sixth S-2A built, and originally came out of the factory registered like all Pitts Specials with a sequential number starting with 80001. Ours being the sixth was N80006. Unfortunately, they didn't number the airplanes sequentially as they were produced, since a special N number did not replace the corresponding 80000 series block. For example, if a custom number like 16 Papa Sugar was put on at the factory, its 80000 number was simply applied to the next airplane,

Sixteen Papa Sugar holds a knife edge over rural New Jersey.

Takeoff in the S-2A is one of the most exhilarating experiences possible.

making the N number useless as an aid in trying to identify an airplane's age or serial number.

Incidentally, the reason we didn't get N6PS (sixth Pitts Special—get it?) was because all of the single-digit numbers with a PS suffix were reserved by the Parks Service. That was twenty years ago and held true until very recently, when we liberated the N6PS number and put it on our plane.

Pitts S-2As came out of the factory equipped with a 200 hp IO-360A1A Lycoming, the exact series of which varied depending on where the airplane fell in the production. The first few planes were equipped with straight IO-360A1A fuel-injected 200 hp Lycomings, but the sumps were modified into the configuration that was eventually recognized as the AEIO-360A1A.

Almost from the beginning of the S-2As, the Hartzell propeller proved to be a problem because it cracked in the shank where the bearing race is located. Art Scholl was supposedly the first to discover this little problem when his prop went into fixed pitch mode and he landed to find one blade with a crack three quarters of the way around the shank. A flurry of ADs (advisories issued by the FAA) came out shortly after that, including various different types of inspections and fixes, one of which glued the bearing race to the propeller to stop it from moving because it was fretting the aluminum, causing stress

risers. It was Scholl's near loss of a propeller blade that caused most Pitts owners to loop stainless-steel control cable around the top edge of the motor mount and the fuselage tubing. This way, in case a propeller blade was thrown, at least the engine wouldn't follow it, and the airplane's balance would allow a controlled crash.

During the early years there were lots of minor teething problems, including ADs requiring the replacement of the top wing main fittings because they bent and cracked under heavy negative g-forces. Doublers were put on the aileron spars to cure a cracking problem at the hinge points, and the horizontal tail got a strut leading from the forward edge of the stabilizer to the bottom of the fuselage. This strut was added because the carry-through tube was cracking as a result of pilots insisting on pushing the airplane into the hangar by picking up on the leading edge of the stab.

One AD that struck a lot of Pitts owners as being a little silly was the replacement of the streamlined slave strut between the ailerons with a round one because a few of the streamlined ones had vibrated, causing flutter in flight. Shortly thereafter an STC (supplemental type certificate) was granted to allow individual airplanes to put the original tubings back on.

Another minor problem surfaced when it was discovered the first ten airplanes had primer on the tubing that

Sixteen Papa Sugar is the sixth S–2A made, rolling out of the factory in August 1971. The author and his partners have owned it since then.

could be scratched off with your fingernail. This was a matter of the factory being sold faulty primer by the vendor. The factory, however, did stand behind the airplanes and paid to have them stripped down to the last nut and bolt, sandblasted and brought back up. Ours was one of those airplanes.

The majority of the airplanes came out of the factory covered in synthetics, notably Dacron and Ceconite. And the very first ones probably had one of the slickest dope finishes ever put on factory-built airplanes. Unfortunately, this also meant they had one of the thickest finishes ever put on a factory-built plane. This thickness caused the finish

to have a tendency to crack or break in stress points and especially around the cockpit where it would get hit by safety belts and elbows. It wasn't a particularly dangerous problem but now that the airplanes have some age on them, those with the original finish find parts of the fuselage finish all too willing to depart the airplane in flight, leaving portions of the Ceconite stark naked on returning to the airport.

As long as we are talking about everything wrong with the airplane, there was also an AD for cracks which developed in the little conning tower that comes up off the control torque tube and holds the stick in the rear pit.

Well, now that I've painted a truly gruesome picture of the airplane, several things should be pointed out, the first of which is that practically all new airplanes go through similar teething processes and almost none of them take

the vigorous beating the first test-tube Pitts did. Second, just about all the problems I've mentioned disappeared somewhere in the first fifty or so airplanes, indicating that the boys at Afton, Wyoming, were listening to the marketplace.

The two-place Pitts Specials are entirely different animals from their little single-place counterparts. They have a different personality, they have a different feel and in reality they have a different mission. They were designed to open up the world of vein-busting, serious acrobatics for a friend, student or local masochist. No, it won't perform with the S-1 Pitts, but it will do lots of things the single-holer won't do, not the least of which is teaching students and opening this world up to friends and lovers.

The first thing a pilot notices when dropping down into the cockpit of a two-hole Pitts is he is going to feel as if he stepped into an open manhole since he

84

sits really deep in the machine with his head barely showing. Even if the pilot is an experienced S-1 driver, he will feel the manhole syndrome, especially if he's sitting in the back seat. If he's sitting in the front seat it's possible to put a cushion under him and approximate the sight window of a single-holer. This is handy when checking out a pilot before he straps on a single-hole Pitts.

The pilot's legs in the back wrap around the front seat, with the feet coming to rest on the rudder pedals just ahead of the hips of the guy in the front seat. This means a pilot will have to keep his eyes open so the front seat parachute doesn't slide one way or the other, crowding his feet or possibly snagging a rudder pedal.

That front seat passenger also eats up a lot of the room behind the rear instrument panel, meaning there really isn't enough depth for many of the instruments. Rather than move the instrument panel toward the pilot which would crowd him substantially, the instrument panel is bent in the middle, which forms facets with the middle one being almost 3 in. forward of the corners of the panel. This allows some of the gauges to be farther forward for the nearsighted pilots.

This lack of instrument panel depth is also one of the primary reasons it is so difficult to adequately instrument the airplane for serious cross-country flying. Even if the panel were large enough to accept a number of extra navigational instruments, it is not deep enough to allow most of them to fit. For that reason, practically every S-2A has a radio mounted on the back of the front seat down between the pilot's legs, although there are a few S-2As out there with radios mounted above the pilot's knees under the right side of the instrument panel. That not only makes it difficult to get in, but also means a pilot with any height, for instance if he's 6 ft., 3 in. or taller, has no place for his knees.

At least four Pitts Specials have been certified for IFR flight, but in these, front seating is very cramped because the bulkhead that forms the back of the seat was bumped out to give the gyros enough depth.

The FAA's "average" pilot, meaning 5 ft., 10 in. and 170 lb., will find the cockpit to be absolutely tailor-fitted for him. It only gets tight if he's out there playing in freezing temperatures and has to wear a snowmobile suit or heavy jackets. As the pilot's size creeps up to 6 ft., however, the cockpit gets increasingly smaller until some of the really big boys will fit in the airplane only if their height is proportional between legs and torso. It is perfectly practical, however, for a pilot as tall as 6 ft., 6 in. to fly the machine with no special modifications to either him or the airplane. A bigger problem exists with pilots under 5 ft., 6 in., because it is difficult for the short guys to have full rudder travel while still getting full back stick for landing—especially in the back seat.

The airplane in general is designed with absolutely no thought toward creature comforts—and that means creatures of *any* kind. So there are lots of naked tubes and rudimentary controls that will make a Spam cam driver think the airplane was built by blacksmiths. In the back seat that's no big deal, since as soon as the engine starts, the inside of the airplane ceases to exist. In the front seat, however, the back angle is several degrees more upright than the back seat because of the need to give the rear instrument panels as much depth as possible. Unfortunately, that means it begins to assume the character of a torture chamber after extended aerobatics in it. A lot of Pitts pilots will cut a foam wedge cushion that is 2 in. at the bottom and tapers to nothing at the top, moving the passenger's butt forward so he won't come out of the airplane bent over like a troll on medicaid.

If what I am describing sounds something less than hospitable, that's because it is. The two-place Pitts Special pampers the pilot less than almost any other airplane in the world. Every driver will invent his own combination of cushions to adjust the seating pattern so he at least is comfortable in that regard. But there's another kind of comfort that comes from the way the airplane fits, once you've found the right cushion combination. This is especially true if the airplane is equipped with a safety belt system such as Jack Hooker's special harnesses which not only combine two belts onto a single pad, but put a tightening ratchet on one of them. With this arrangement, the pilot can strap himself into the airplane so tight he can literally cut himself in half. The combination of feeling securely nailed to the seat and surrounded by a form-fitting, businesslike cockpit gives tremendous confidence from the beginning . . . once the intimidation begins to fade.

Even before the key is twisted to bring the Lycoming into action, just touching the stick should give the intuitive pilot a feeling of what he is about to experience. If, when sitting on the ground, the stick is pushed to one side with a finger, it moves with absolutely no pressure and then smoothly rights itself in a gentle rocking motion as the ailerons return to neutral. In other words, there is not an eyelash worth of friction anywhere in the system, since every bolt goes through a ball bearing and every one of the bearings is perfectly aligned. It is undoubtedly the most friction-free control system ever installed in a factory-built airplane.

Saddling up is an experience in itself. As I swing my right leg over the cockpit side and then step in to let my butt slide down the back of the seat, it is as if I am performing some nearly obscene rite of passage as the airplane comes up to take me into its fold. I am very conscious of watching myself strap into the machine, and I feel the top edge of the instrument panel and the edges of the sheet metal at the side defining my environment. If someone is helping me with the shoulder harness and adjusting my parachute, the feeling is one of a squire fitting me with medieval body armor. As he slides the canopy closed, it is as if he is putting the helmet down over my head to complete the transformation from civilian to Pitts warrior.

Starting the S-2 when it's cold means toggling the primer switch *up* with the mixture forward and counting to eight seconds. The mixture comes out lean again, the throttle is cracked and the inevitable shout of "Clear!" makes sure no one is out there near the Hartzell Cuisinart. Then the key is turned to the start position and one of the neatest sounds in aviation is created. On the fourth or fifth blade the air-fuel mixture will be just right and a couple of coughs indicate it is time to slide the mixture control forward, letting the fuel injection spray fuel into the cylinders in the proper sequence. Almost instantly there is an explosion of noise that can be felt under my feet where the exhaust stacks start belching out their unmuffled bark.

Sixteen Papa Sugar turns through base leg during approach to its home base in Andover, New Jersey.

The airplane has no carburetor heat and even less cabin heat, and there is no need for mufflers of any kind. These little straight pipes have a hard, sharp, authoritative sound that brings a smile to my face almost every time.

As the engine catches and the throttle is brought forward to bring it into a fast idle, the airplane will rock gently from left to right and the noise inside the cockpit is considerable without a headset or earplugs. In fact, from the outside the airplane on takeoff generates 92 decibels (dB). Inside the cockpit, in

excess of 125 dB is generated on takeoff, nearly 30 dB above the level where the all-knowing, all-powerful Occupational Safety and Hazards Administration (OSHA) says permanent hearing damage can be caused. I can't attest to that, but I can say there is some serious noise going on in there.

On the ground, all S-2 series Pitts are noticeably blinder than their single-place cousins and it is necessary to do plenty of S-turns to see around the nose. There have been any number of incidents where pilots have nearly taxied into airplanes and trucks simply because they didn't see them. Doing the S-turns is no big deal, since the tail wheel is nicely steerable; however, the first time out it is a good idea to keep your feet up on the pedals so you can get to

the toe brakes if you need them. This is because even the gentlest S-turn can lead into a fairly healthy swerve which could cause the tail wheel to unlock and go into full swivel. This could easily cause a ground loop right there on the taxiway, which wouldn't put a mark on the airplane but it would certainly pound your ego into the dirt.

During the run-up, the crude little prop control protruding from the upper left-hand corner of the instrument panel is brought back a couple of times to make sure the prop is indeed working,

Pitts pilots know total freedom when flying any model.

The flight deck of a factory new S-2A fits like a glove—a tight one.

and the equally crude trim tab down by your left knee is adjusted so it is in the middle of the takeoff indication. Usually, I'll look back over my left shoulder at the horizontal tail and make sure the trim

tab is streamlined with the elevator. Then I roll out into the middle of the runway and do what is probably the neatest thing I'm going to do all day—a Pitts takeoff.

Since the propeller is so big in relation to the size of the airplane and there is so much power available, the throttle is treated gently on take off. View it as if it

is a dynamite plunger—since that's exactly what it could be. It is also not a red-hot idea to immediately force the tail off the ground like a 65 hp Aeronca Champ, since that's inviting an amazingly immediate left turn caused by the precession of the heavy propeller. The name of the game is smooth and gentle.

As the power is smoothly fed in I keep the stick sucked into my lap until the throttle hits the stop, by which time we are streaking down the runway fast enough the rudder has lots of wind in it so I gently bring the tail up into a slightly tail-low attitude. At this point a lot of pilots will pick the tail up too high, thinking they should be able to see over the nose. That's not only bad technique, it's a lousy idea since it puts the airplane at a slightly negative angle of attack forcing you to drop the tail to help it fly off. I much prefer to bring it up into a less-than-level attitude, keeping the tail just slightly low. As the speed builds, I gradually increase back pressure on the stick, helping the plane find a speed it likes so it can fly off on its own.

During that takeoff run I like to have students notice what they are seeing or, better yet, what they are not seeing. On a narrow runway, they are probably not seeing much of anything but a tiny sliver in that aforementioned triangle in each corner. On a wider runway, maybe they'll see a little more, but not much. But the objective during this streaking trip down the runway is to keep the airplane straight with gentle pressure one way or the other until it lifts off.

Once it's off the ground, there is no need to drop the nose and wait for speed to build, as is common with most airplanes because by the time you get your wits reassembled, the airplane is already passing through 80 or 85 mph. In fact, less attentive pilots may find themselves doing 80 or 85 mph while still on the ground. This too is lousy technique, since those little 500x5 wheels are spinning like turbines at that speed.

As a normal rule, the nose doesn't have to be lowered at all to get up to best rate-of-climb speed of 94 mph. In fact, since the nose is so steep in that kind of situation, many pilots climb at 100 or 105 just so the airplane isn't at such a ridiculous angle. This is especially true if it's cold and the air is dense. At 94 mph the book says you'll get nearly 1,800 ft. per minute (fpm) climb with two people on board and an aerobatic gross weight of 1,500 lb. (the airplane's empty weight is about 1,050 lb.), but that depends on several factors, temperature and load being the primary ones.

For me, one of life's true delights is sneaking out to the airport on a cold winter day and shooting touch-and-goes. I usually have an attentive crowd assembled in the windows of the small terminal lobby by the second takeoff. In below-freezing temperatures, I have to forcibly bring the nose up after takeoff to maintain 94 mph, which results in an awesome takeoff and climb angle. Folks seem to get a kick out of watching it. None gets as much of a kick as I do. With the air that fat and solid, it is as if that big prop is physically chopping chunks out of it and flinging them back to propel me up and away. The airplane feels like it is sitting in the air as solid as if it were on the ramp, except the ramp and all real estate in sight is falling behind as if part of a *Star Wars* special effect. It is a truly intoxicating experience.

Very few airplanes the normal pilot flies will be as obviously affected by density altitude as are the Pitts Specials—all of them. The S-2 models, however, have the option of stuffing another 170 lb. in the front pit, and that will produce a more noticeable change than simply temperature on a single-place airplane. The difference between an S-2A with just the pilot on board and one loaded up is dramatic. It's not unusual to see solo climb rates in the 2,500 fpm category during the winter months, and see that rate drop to 1,600 or 1,700 fpm during the summer. On a 95 deg. day, with a passenger on board, I feel lucky to get 1,000 fpm. Exactly the reverse is true, of course, on landings since fat air wants the airplane to stay up longer and hot, thin air wants to make it fall out of the air. Don't take this to mean the airplane ever becomes anything less than an amazing performer. In fact, in the worst situation it will blow anything made in Wichita into the weeds.

Compared to the S-1 airplanes, the ailerons and roll rate of the S-2s all seem much slower, especially if they are not equipped with shovels. Shovels attach to the spars on the lower ailerons and drop down 8 in. or so, mounting a flat plate that is located ahead of the aileron's hinge point and a sizeable distance below it. This creates an aerodynamic load

which helps move the ailerons when they're deflected. The aileron forces drop drastically, as much as 50 percent. This also supposedly helps the roll rate, although there are other little tricks that can be done that help even more.

Besides shovels, the most important thing that can be done to the airplane to

Accommodations in front are Spartan.

improve its roll performance is to seal the gaps in the ailerons with a flexible seal made of fabric or tape. This stops the flow of air from the high-pressure side of the wing through the gap in front

make the sport pilot feel as if he has his hands on a mini-missile that is hooked directly to his brain. Every aspect of the controls are going to be much, much lighter and quicker than anything he has experienced.

In the air, the S–2A has a feel of its own—it feels as if the plane is physically pushing against the air and moving in a precise, measured manner, a feeling best described as dense. It is an intangible feeling, but is nonetheless a feeling many pilots comment on after climbing out of the airplane for the first time. Many custom-built airplanes can outperform the Pitts in terms of roll rate and climb, but almost none have the Pitts' peculiar feeling. Even the seemingly similar Christen Eagles, a near clone, lack it. The S–2A is unique among airplanes.

Sitting in that form-fitting cockpit, the plexiglass canopy keeping the elements out and the noise in, is the closest most will come to knowing true three-dimensional freedom. It's not a matter of moving the airplane, it is a matter of moving myself, with the airplane so firmly attached to me it has no choice but to follow. With the 204 mph readline so high and the 9–g limit so far away, there are few limitations to what I can ask my body to do.

If I'm just up goofing off, chances are, as I come into the practice area, I'm already emotionally wired just from the short flight over from the airport. Knowing what's coming, I reach down and put another two extra clicks into Hooker's fabulous seatbelt ratchet. The airplane and I are welded together and are on the edge of an emotional high. It takes nothing to push us over the edge.

There's no telling what will happen first, but it will involve gaining speed and might mean pulling the nose up high to roll over on my back, letting the nose fall through the horizon in a screaming, speed-building split S. Chances are, until I'm inverted and the nose is well on its way down, I haven't made up my mind what will come next. If I'm in a vertical mood, my right hand knows just how hard to pull to make the speed hit 185 mph as I pull out level. Then, the stick comes back firmly and I feel the g-force lay its heavy hand on me during the pull into the vertical. My head is pivoting back and forth from wing tip to wing tip, ensuring that the line is at right angles

The S–2A, like all Pitts Specials, lets the pilot catch himself coming or going.

of the aileron, increasing its effectiveness and making a noticeable difference in roll acceleration: when the stick is moved, the airplane reacts faster.

If the elevators and rudders are gap sealed as well, the plane's overall control effectiveness is increased a great deal. But bear in mind this is like improving the speed of a rifle bullet by a small margin. How fast is fast and how good is good? Even the absolutely dead stock S–2A Pitts with no seals of any kind will to the world before my right hand searches for the left side of the cockpit, the stick moving with it, and the airplane obediently pivoting on its axis. If the wing tip rips across the horizon without straying up or down, I grin. I grimace if it wanders off.

As I return to the point where I began, the ailerons come out and the airplane halts. It's just hanging there, the engine thundering away but to me unheard, as I feel the airplane speaking to me. Then, without knowing why, my left foot instinctively senses the right moment and hammers the rudder to the floor pivoting the airplane on its rudderpost to leave me pointing straight down. Sometimes, not always, I'll kill the power as the airplane approaches the down vertical. If I do, it seems to hang there, gravity not having taken hold, as if undecided whether the plane should come down or not. The moment seems endless. The airplane is quiet. The world is directly ahead. Straight down is my direction, but the airplane plays with gravity for a while, giving me time to reflect on my next move.

As the speed begins to build, the ailerons go in again, and the landscape ahead rotates halfway around. As I stop it, my right hand reaches forward, pushing the stick firmly under the instrument panel and my left hand moves the throttle forward. The belts bite into my hips, as the airplane curls under into inverted flight. I grin slightly. I know I'll come home from this trip with black and blue marks on my thighs, the badges of the negative g pilot. We call them Pitts hickies.

As much as I love the aerobatics, it's landing the Pitts that I love the most. Maybe its the challenge, knowing only one out of four or five will be smooth no-bouncers. Maybe it is simply the challenge of not making a fool of myself. After twenty years in the same cockpit, I still have no idea how the landing will turn out.

On downwind, I kill the power, bring that crude little trim lever up and slide the prop control forward. The airplane slows quickly and I drop the nose to what I know will yield 85 mph. That's the last time I check the air speed. From that point on, I'm concentrating on the runway, trying to fly an invisible path to a point near its end. I know, if I control the attitude and the path just right, I'll

get a good landing. Maybe. Just as surely, I know if I let the airplane get low, or fast, or do anything to wander outside the narrow, unseen tube that defines the proper profile, I'll get a bad landing. Fly a bad approach and be guaranteed a bad landing. It's as simple as that.

Almost as soon as the power is back and the speed bled off, I drop the wing and start arching toward the runway. The guessing game has begun. I want to fly an uninterrupted 180 deg. turn that will end with me gently touching down on the runway in a three-point position. But, it won't happen that way and I know it. There are too many variables in between that must be compensated for.

The turning approach is gauged so the corner of the runway always sits just barely in sight on the side of the nose. If I were to fly a normal rectangular approach, like a Spam-can pilot, the second I rolled out on final, I would be unable to see any of the runway. Fly a normal approach, and a 3,000 ft. runway simply disappears from the time the airplane turns final until it is within 15 ft. of the ground. In fact, on a straight-in final, the airplane is so blind, there's a good possibility the hangars and other major portions of the airport will disappear. In a two-hole Pitts, straight finals are avoided at all possible costs.

As I turn toward the runway, I try to visualize where the airplane is going to wind up versus where I've already visualized it is supposed to go. I try to picture that small invisible pathway that defines a good approach, and work to keep the airplane on it. I know I don't want to be low, since adding power on short final is another way to almost guarantee a bad landing. Even the slightest amount of power upsets the profile, when close to the runway. The nose comes up and the speed changes. The balance is gone and so is the profile.

Given a choice, I like to be a little high, since losing altitude is not one of the Pitts' weak points. Simply reducing the power starts it downhill at a breathtaking rate. However, it is absolutely the finest slipping airplane ever designed, so the pilot has total control over how much rate of descent he or she wants. Cross control a little, and the nose slides sideways and the airplane comes down just a little faster. Put the rudder hard in one corner and the stick in the other and the airplane develops frightening rates of descent. It simply falls out of the sky as if it were dynamic, rather than aerodynamic. However, no matter how fast it is coming down, the second the controls are neutralized, the increase in rate of descent stops. It just stops. It's almost magical. For that reason alone, most of the turning approaches I fly will be high and tight, with a touch of slip thrown in for total control and added visibility.

As the ground comes up to meet me, I remind myself to put my head back and concentrate on both sides of the runway at one time. This is critical on narrow runways, since they don't show up until it's almost too late to make any corrections. By watching out of both sides at the same time, it's easy: If one side pops into view several seconds ahead of the other, that's a sure indication the airplane isn't on the centerline. Finding the center of a narrow runway is the hardest, and most critical, part of landing the airplane the first few times. After that, it is still difficult, but manageable.

Once the ground is in view, my job is to gauge my height above it and hold the airplane in the proper attitude, so all three wheels touch at one time. All but the last fifty S-2As produced sit on the ground at a 9 deg. deck angle, which is far short of its actual stall angle. In fact, a full-stall landing will put the tailwheel on the ground while the main gear is still 6 in. in the air. For that reason, landing the plane isn't a simple matter of keeping it in the air as long as it will fly. As with everything else in its personality, it is different from most airplanes. It must be held at just the right attitude. If the mains touch first, the airplane will skip and hop and generally embarrass me. If the tail touches first, it will plop down and generally embarrass me. If all three smoothly touch, it will generally surprise me.

Once on the ground, my toes caress the brakes, slowing the airplane without asking too much of it. On grass, the plane will track more or less straight ahead, as the drag on the tailwheel acts as a tailskid to keep the tail in back, where it belongs. If it is pavement, however, I'm ready for the small collection of mini-swerves the airplane will throw at me. I know it won't roll straight and it will expect my feet to do a little tap-dance, keeping it in line. As it slows to a halt, and I turn onto a taxiway, I know the flight isn't over, but the major challenges have been met.

The best part of the flight is swinging the airplane around in front of the hangar, pulling the mixture and sliding the canopy back. As I yank the straps loose and throw them over the sides of the cockpit, it is as if part of my soul has had an energy transplant. I feel revitalized. I'm a new person who is ready to take on all comers. I feel like shouting at the world. Sometimes I do. Those at the airport have come to expect it. They don't know the joy of flying a Pitts, but they know the joy of the Pitts pilot. This one can't keep it hidden.

In the S-2A, two can play.

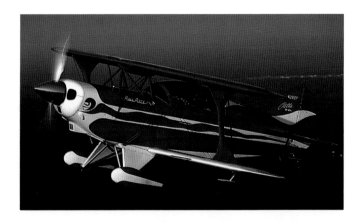

Factory-Built Birds
The S-1S, S-1T, S-2A, S-2B and S-2S

The year 1971 was a heady year for American aerobatics. Not only were American pilots beginning to bring gold home from the world contests, but they now had a factory devoted entirely to the production of advanced aerobatic airplanes. Granted, the airplane was the S-2A, a two-placed trainer that couldn't compete in world competition, but it was a start. A serious start.

For the first several years, Anderson and Pitts were hard pressed to keep ahead of their orders. In fact, there was a period at the beginning where, if you ordered an S-2A, the best they could promise was a year or more delivery. Then, they began to get requests for an airplane that was a world beater and still certified. What the market was asking for was a certified S-1S.

The market is never satisfied and began asking for more horses in the S-2A. They began making noises in other areas, and before long it was obvious the Pitts factory was going to have an ongoing product line.

S-2A

The S-2A remained essentially unchanged from its introduction in May

An S–1T (foreground) and an S–2A fly formation.

The Afton, Wyoming, Pitts Factory would paint your Pitts any way you wanted, and many, like this S–2A, were produced with custom paint schemes.

1971 to when production ceased in March 1982. The total number of airplanes built was 272, less than half of which are still in the country. From the beginning, approximately 30 percent of production went overseas and a sizeable number of used S-2As have been exported since.

The only major changes to the airplanes included the introduction of a 2 in. wider fuselage in November 1979, beginning with airplane serial number 2206. At the same time a longer landing gear was incorporated that raised the deck angle from 9 deg. to 12 deg. This was done to reduce the landing speed.

Symmetrical ailerons were first tried on serial number 2219, and aerodynamic shovels or spades were introduced as part of that change. These ailerons were introduced on the production line beginning with serial number 2231 in June of 1980.

In determining changes, serial numbers should be used rather than production dates. Some aircraft came out of the factory out of sequence because they were held in the factory longer for special equipment or paint jobs, which allowed later aircraft with higher serial numbers to be delivered ahead of them.

Throughout the years numerous minor improvements were incorporated, including the horizontal stabilizer strut, two-piece nose bowl, metal landing gear V covers and other smaller items. How-

ever most of the basic airframe was unchanged.

S-1S

The factory S-1S took up where the homebuilt models left off. There are practically no major structural or aerodynamic differences between a homebuilt S-1S built exactly to plans and one built in the factory. However, it should be noted that prior to the factory's introduction of the S-1S in kit form, theoretically there are no such things as S-1Ss, built in their entirety by the builder.

The factory began building certified S-1Ss in August of 1971, barely two years after the S-2A line started up. The airplanes were certified complete with an electrical system which gave the planes self-start capabilities. However, since the machines were designed for competition where weight is the mortal enemy, many airplanes were ordered without the electrical system. Most factory S-1Ss came with the 180 hp Lycoming.

Other options included a center section tank in the top wing for smoke oil, a lexan, see-through belly and other minor changes. For that reason, S-1Ss will be seen with empty weights varying as much as 75 lb., and with a wide variety of equipment and paint jobs.

The total number of S-1Ss produced by the factory was sixty-four before production was halted in September

1981 in favor of the updated single-place, the S–1T. There were, however, a large number of S–1S kits sold, but even the factory is unsure of the exact number. The reason for the ambiguity is some of the aircraft went out the door as kits, while others were bought as kits and then finished by the factory. A few of these were certified. The best estimate is that in excess of 100 S–1S kits were built and sold.

The S–2S is considered by some to be one of the best Pitts variants and was eventually to give birth to the S–2B two place.

Because of liability concerns, shortly after the factory was bought by Christen Industries, the practice of selling kits and supporting those in the field was halted. However, as this book was going to print, it appeared the factory would once again support those airplanes being built by making parts available. Kits, however, will not be available.

S–1T

By the early 1980s, the Pitts was seeing hard-nosed competition from a number of monoplane designs, most notably the Stephens Akro as modified into the Laser 200 by Leo Loudenslager. In an effort to build an even higher per-

The S–2S and S–2B differ from most S–2As not only because of the larger engine, but also because they used the wider fuselage and longer landing gear of the very last S–2As.

Next page
The S–2S is capable of speeds approaching 190 mph and carries an extra fuel tank for range.

formance single-place machine, the factory modified the S–1S design into the S–1T.

The T models' most notable change was the installation of a Hartzell con-

Pitts Special S-2S

The single-place S-2S has the greatest vertical penetration and highest speed of any aircraft in the Pitts Special line. The S-2S is powered by a 260 hp AVCO/Lycoming AEIO-540-D4A5 engine with a Hartzell constant-speed propeller. With its 35 gallon fuel tank, the cruising range of the S-2S is over 400 miles with a 30 minute reserve.

The standard S-2S includes a fixed windscreen with sliding canopy and the stripe-and-diamond paint design, as shown. Additional standard features include altimeter, airspeed indicator, compass, accelerometer (g meter), recording tachometer, oil temperature and pressure gage, fuel flow and manifold pressure gage, cylinder-head temperature (CHT) gage, stall-warning system, and electrical system including gel-cell battery. A Christen seat belt system with secondary lap belt is supplied, and the engine is equipped with a Christen inverted oil system.

The relatively large size of the aircraft makes the S-2S ideal for demonstration flying at airshows, and it is easily judged from the ground at aerobatic contests. In addition, the relatively large cockpit is capable of accommodating pilots up to 6 ft. 4 in. height and 250 lb.

S–2S specifications courtesy of Christen Industries.

stant-speed propeller. This was done to allow the engine to continue developing full power even when the aircraft has slowed almost to a halt, as at the top of a vertical roll.

Coupled with the propeller is the 200 hp, IO-360 Lycoming, as used in the S-2A. Theoretically, the higher horsepower and the constant-speed prop would have increased the airplane's performance. Unfortunately, these changes, along with the inclusion of an electrical system, raise the T's empty weight to 830 lb., which handicaps its performance, when compared to a lightweight S-1S. It is, however, a very civilized airplane that lets its pilot go anywhere, at any time,

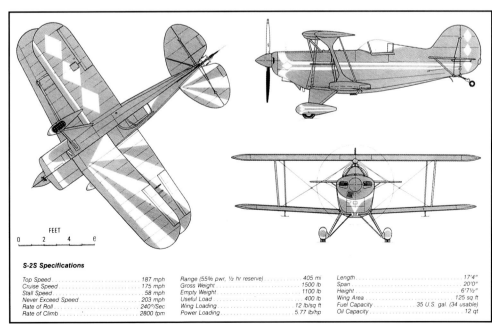

FEET
0   2   4   6

S-2S Specifications

Top Speed	187 mph	Range (55% pwr, ½ hr reserve)	405 mi	Length	17'4"
Cruise Speed	175 mph	Gross Weight	1500 lb	Span	20'0"
Stall Speed	58 mph	Empty Weight	1100 lb	Height	6'7½"
Never Exceed Speed	203 mph	Useful Load	400 lb	Wing Area	125 sq ft
Rate of Roll	240°/Sec	Wing Loading	12 lb/sq ft	Fuel Capacity	35 U.S. gal. (34 usable)
Rate of Climb	2800 fpm	Power Loading	5.77 lb/hp	Oil Capacity	12 qt

Pitts Special S-2B

The S-2B is the outstanding two-place version of the Pitts Special. Equipped with a 260 hp engine, this aircraft has the unique capability of true unlimited-class performance, even while carrying both a pilot and passenger. The Pitts Special S-2B is the preferred aircraft for dual instruction for all aerobatic maneuvers. In addition it is a suitable aircraft for limited cross-country flying. Like the S-2S, the S-2B is powered by a 260 hp AVCO/Lycoming AEIO-540-D4A5 engine with a Hartzell constant-speed propeller.

The standard S-2B includes a fixed windscreen with jettisonable canopy and the stripe-and-diamond paint design, as shown. Additional standard features include altimeter, airspeed indicator, compass, accelerometer (g meter), recording tachometer, oil temperature and pressure gage, fuel flow and manifold pressure gage, cylinder-head temperature (CHT) gage, stall warning system, and full electrical system including gel-cell battery. Basic engine and flight instruments are duplicated in both cockpits. A Christen seat belt system is supplied for both cockpits with secondary lap belt provided for the rear cockpit, and the engine is equipped with a Christen inverted oil system.

The S-2 and S-2A are earlier versions of this aircraft type. They are both similar in appearance to the S-2B, but are supplied with smaller engines. Open cockpits with fixed windscreens are standard. The S-2, which has never been available as a production aircraft, is much lighter because it is equipped with a 180 hp engine and fixed-pitch propeller. The S-2A has a slightly heavier feel than the S-2 but is somewhat higher powered; it is equipped with a 200 hp engine and constant-speed propeller.

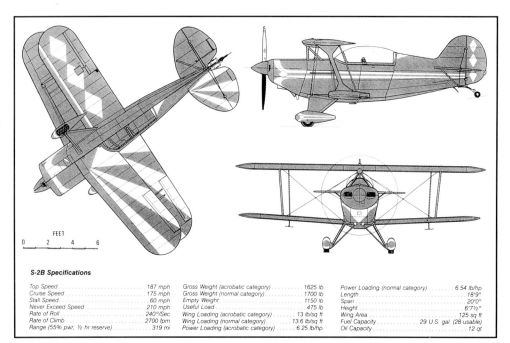

S-2B specifications, courtesy of Christen Industries.

without worrying about the availability of an experienced person to prop it.

The T model went into production in May of 1981 and is still available on a special-order basis. More than fifty of the aircraft have been delivered.

S-2S

The S-2S was a logical progression for the S-2A: The larger airframe of the S-2A seemed ready made for hanging on a six-cylinder, 260 hp Lycoming, and it wasn't long before customers began asking for such an airplane.

Initially, the S-2S was an experimental machine, since it was a modified A model and wasn't certified in the standard category. The first series of S-2Ss

S-2B Specifications

Top Speed	187 mph	Gross Weight (acrobatic category)	1625 lb	Power Loading (normal category)	6.54 lb/hp
Cruise Speed	175 mph	Gross Weight (normal category)	1700 lb	Length	18'9"
Stall Speed	60 mph	Empty Weight	1150 lb	Span	20'0"
Never Exceed Speed	210 mph	Useful Load	475 lb	Height	6'7½"
Rate of Roll	240°/Sec	Wing Loading (acrobatic category)	13 lb/sq ft	Wing Area	125 sq ft
Rate of Climb	2700 fpm	Wing Loading (normal category)	13.6 lb/sq ft	Fuel Capacity	29 U.S. gal. (28 usable)
Range (55% pwr, ½ hr reserve)	319 mi	Power Loading (acrobatic category)	6.25 lb/hp	Oil Capacity	12 qt

The final variant of the single-place Pitts is the S–1T, with the 200 hp engine and constant speed prop of the S–2A. Note the aileron shovels.

were referred to as 1000 series and they were all certified in the Exhibition sub-category of the Experimental category. What that meant was they didn't have to conform to the FAA rules regarding certification, and owners could have them built anyway they wanted. For that reason, 1000 series S-2Ss are seen with both spring steel and V landing gear. As the airplane was being groomed for certification, the spring gear was dropped because it couldn't be made to pass the

S–1T specifications, courtesy of Christen Industries.

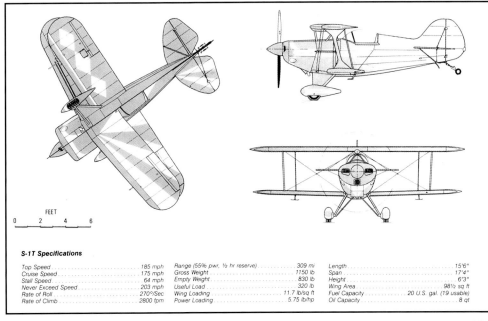

S-1T Specifications

Top Speed .... 185 mph	Range (55% pwr, ½ hr reserve) .... 309 mi	Length .... 15'6"
Cruise Speed .... 175 mph	Gross Weight .... 1150 lb	Span .... 17'4"
Stall Speed .... 64 mph	Empty Weight .... 830 lb	Height .... 6'3"
Never Exceed Speed .... 203 mph	Useful Load .... 320 lb	Wing Area .... 98½ sq ft
Rate of Roll .... 270°/Sec	Wing Loading .... 11.7 lb/sq ft	Fuel Capacity .... 20 U.S. gal. (19 usable)
Rate of Climb .... 2800 fpm	Power Loading .... 5.75 lb/hp	Oil Capacity .... 8 qt

FAA drop test requirements without being ridiculously heavy.

The 1000 series airplanes served as test beds for several ideas, including the 260 hp installation and symmetrical ailerons. Eventually, all of these modifications wound up on the certified 3000 series S–2S, as well as the S–2B. The symmetrical ailerons soon found their way into the last of the S–2A production line and continues on the S–1T.

A special-order machine, the S–2S has the forward cockpit of the S–2A/B removed and an extra fuel tank inserted for additional range, always a problem with the little Pitts Specials. There is never much over two hours' fuel on board without extra tanks.

More than thirty S–2Ss have been built, and most pilots consider the S–2S to be the best performing, best balanced Pitts built, which is saying a lot. The reason more haven't been built undoubtedly has to do with its single seat configuration and general similarity to the S–2B, which offers a second seat for greater utility.

S–2B

Building on the experience with the S–2S and the success of the S–2A, the factory decided to totally update and redesign the S–2A into a different aircraft. Using the latest A airframe with the wider fuselage, longer landing gear and symmetrical ailerons, the forward bays and cabane strut were reshaped and strengthened to accept a 260 hp, IO–540 Lycoming while still retaining the front cockpit. Both cockpits are enclosed in the long, swingover bubble canopy that was an S–2A option.

The S–2B completely replaced the S–2A on the production line in December of 1982 because the final costs of the two airplanes were so close, the market for the smaller engined airplane disappeared the instant the B was introduced. With a 30 percent increase in power and a weight jump to 1150 lb., the S–2 completely changed character. The S–2B's cruise speed was higher than the top speed of the A and its climb rate went from approximately 1,700 fpm to over 2,700 fpm.

Although much heavier, the S–1T offered cross country convenience with an electrical system and higher speeds. Notice the pressure cowl.

PITTS SPECIAL S-2A

Illustration courtesy of Air Progress.

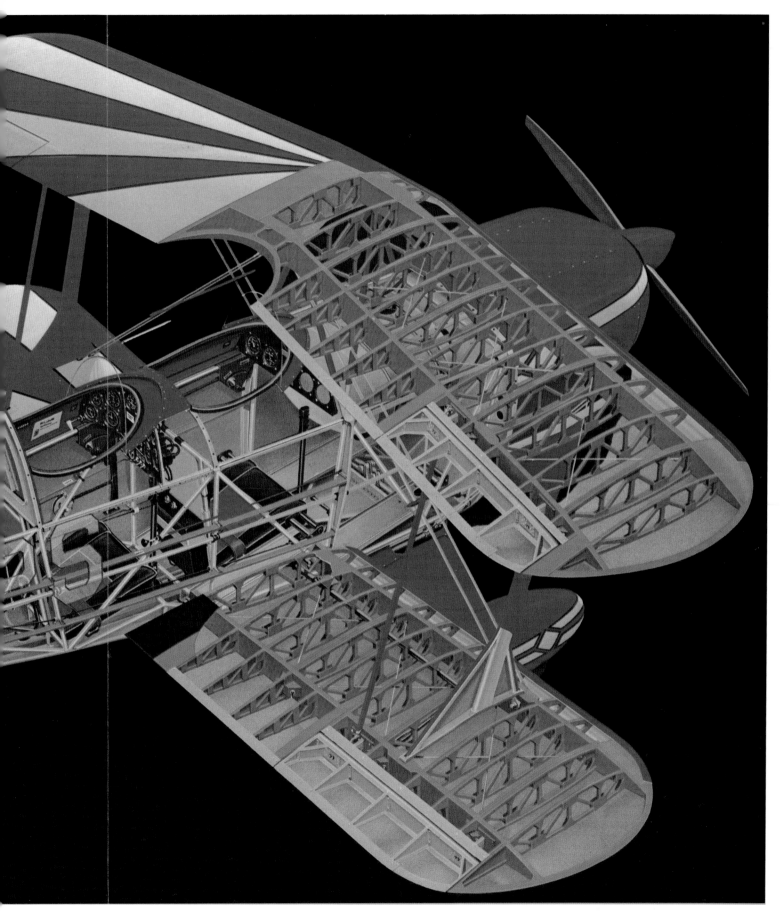

Previous page
The S–2A (background) has been out of production since March of 1982, but the S–1T has continued into the 1990s.

Family resemblance between the S–1T and S–2A is undeniable.

Next page
The symmetrical wings make the S–1T, and all roundwing Pitts, ideal for outside or inverted maneuvers.

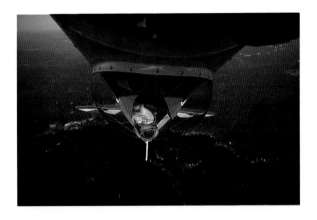

Buying a Pitts
Trading Greenbacks for Grins

There are only two possible ways to deal with an uncontrollable urge to jump into the wild and wooly world of the Pitts Special. The first is to sweep out the garage, buy a welding torch and spend the next three years of your life cutting and whittling. The other obvious route is to trade greenbacks for grins which, although much more immediate than actually building a Pitts, nonetheless should be preceded by a lot of evaluating and even more planning before asking the bank to put all your available dollars in a brown paper bag.

The necessity of planning and evaluation must not be underestimated when buying an airplane such as a Pitts Special. Although the same process applies to buying any airplane, when a machine has as much performance and high-profile activity attached to it as a Pitts, this planning and evaluation process becomes even more critical.

Often there is a general assumption the only evaluation to be done is hardware oriented, for example, kicking the

The Ultimate wings on this S–1C are easily identified by their square tips, as contrasted with the S–1P in the background.

On the top of a loop, akro pilot Glenn Giere surveys his world from the back of an S–2A.

tires and peeking at the bolts to make sure a given plane is airworthy. Actually, giving the airplane a close physical examination is the last and the easiest step in the process, since the pilot should spend at least as much time evaluating himself and his lifestyle before tossing a biplane into the middle of it. If the personal evaluation process isn't complete, the result can be like throwing a grenade into a chicken coop.

Only after the individual has done a first-class evaluation of himself and his situation should he start looking at specific airplanes as a place to put his bucks.

Evaluating Yourself as a Pilot

Rule number one before buying a Pitts: If you haven't flown a Pitts, go buy some time in one. That flight will either turn you off or set your mind on fire. Either way, it's important.

An airplane such as a Pitts Special absolutely defines high performance and, therefore, requires the pilot to recognize the need for good stick and rudder capabilities. No, better than that, the pilot must recognize that he is climbing into one of the most demanding airplanes ever built and not only will it continually test his skill, even after he has learned to fly the airplane, but it will present a learning challenge that he may or may not be willing to grow to meet.

There is one basic fact of aviation that must be remembered: Any pilot at any level of experience can learn to fly any airplane. There are no airplanes, including Pitts Specials, that require super-human talent to fly. However, the Pitts does require the pilot be willing to carefully evaluate his own past experience and, if that appears to be lacking, he must be willing to go find the appropriate training in the front seat of an S–2A or B.

As we've said before regardless of the number of hours a pilot has logged and of the type of airplanes he has flown, he can count on his first flight in a Pitts Special, whether a single-place or two-place, to be memorable. More than any other, this takeoff and landing will require his feet and hands be talking to each other and his brain orchestrating the whole thing at a rate never experienced before. Again, this doesn't mean you have to be superhuman, but it does mean you have to be willing to rise to the task. In some cases this may mean only a familiarization flight or a couple of hours; in other, more extreme cases, twenty hours will be spent working up a sweat in the front seat of a two-holer.

Evaluating Your Lifestyle

Before any airplane is tossed into a person's life it is generally helpful if that person will turn around and look at

Ellen Dean of the 1990 US aerobatic team flies in her modified Pitts. Many competition aircraft are heavily modified. Before you purchase a modified Pitts, examine the modifications to ensure that they do not make the aircraft unsafe.

himself and his lifestyle and determine exactly how that plane will to fit into it. That means evaluating not only where he lives but also *how* he lives and how much time he will have to devote to that airplane.

Certainly, an airplane such as a Pitts does not require the amount of free time an airplane designed for cross-country does. An airplane like a Pitts often makes a lot more sense to most people's lifestyle simply because, as a sport airplane they are forty-five-minute machines: you can dash out to the airport, jump into it, spend forty-five minutes

having a good time and then go on with the rest of your life. This is not normally the case with an airplane that is meant to take you places. On a typical Saturday morning, a Pitts will fit very nicely between taking Jennifer to the dentist and picking up Scott at the soccer game.

The Pitts Special, regardless of the model, is also an airplane that does not require nearly the maintenance or support of a normal airplane. It has no complex systems and is nearly as simple as a Piper Cub to keep up. It does, however, require that the airplane be hangared, which raises the question of finances. Buying the airplane is one thing. Being able to support it is something entirely different. In some parts of the country that is not a major factor since hangars may rent for $40 or $50 a month. However, in metropolitan areas the cost of the hangar may rival the cost of a two-bedroom apartment. The good

news is Pitts Specials are such tiny airplanes, they can almost always be squeezed in amongst the bigger machines and are usually charged a much smaller hangar rent.

Money is one aspect of aviation we would just as soon not have to think about. However, before jumping into buying a Pitts, it's a good idea to carefully consider your finances, since the world of the Pitts Special stretches from thirty-year-old, homebuilt, 125 hp flatwings to brand-new S–2Bs with state-of-the-art navigational packages. Again, the good news is that the range goes from less than the price of a new Chevy to a middle-of-the-line Ferrari, with lots of stops in between. This enables the individual to tailor the financial burden to what he can handle. The financial hurdle may be the deciding factor when it comes to single-place versus two-place, since the cost of a really fine 150

hp flatwing may be only a third the price of a used S–2A which in turn is half the price of a new S–2B. On the other hand, a good competition quality, homebuilt S–1S or competition modified S–1S is going to run the price of a next-to-the-bottom-line Mercedes, or half the price of a new S–2B.

In 1990 dollars, a good flatwing should fall in the $12,000 to $18,000 bracket, S–1Ss would be in the $30,000 to $40,000 range, S–2As in the $45,000 to $60,000 range and, at this writing, S–2Bs were running $86,000 new. In other words, from a financial point of view, the range is so wide that anyone who can afford to buy and keep up a good boat can jump into a Pitts Special.

Part of the maintenance cost of an airplane is taking care of the small things that happen from time to time, but which will sometimes require that the local A&P jump on the airplane to fix this or that. This is especially true of the factory-built airplanes, since there isn't a whole lot on the machine the individual can work on legally. Homebuilts, on the other hand, can be worked on by the owner, which is a good thing because, depending on the quality of airplane purchased, homebuilts can often require a little more tinkering to keep them running and flying properly. Either airplane, factory or homebuilt, does require a certain understanding of the characteristics of rag and tube airplanes. This is another way of saying, it helps if you have a local mechanic who has some background in this area, since many of the newer Spam-can mechanics don't know anything about tubing and fabric.

It should come as no secret that Pitts Specials, in general, are something of a special-use airplane. Yes, they can be used to go see Aunt Minnie or to hop over to get one of the famous $30 hamburgers at the next airport, but primar-

ily they are a special-use airplane. They are made to go out and convert fuel into high g's and big grins. Unfortunately, to do that requires something no other aircraft requires, and that is airspace in which it is legal to do aerobatics. This means the airplane must be a minimum of five miles off the centerline of any airway, and 1,500 ft. minimum over an unpopulated area. If the airplane is based eighty miles due west of Omaha, that shouldn't be a big deal. But for those pilots in metropolitan areas, that could be more of a problem than they imagine because airways seem to be everywhere. At the very least, finding aerobatic space would have a bearing on which airport the plane is to be based.

Picking Your Airport

An entire book could be written about the subject of airports that are Pitts-friendly and those that are not. It should be fairly obvious that Pitts Specials are not mainstream airplanes and that those pilots and practitioners of the general aviation art who *are* mainstream (Spam-can jockeys) tend to look at aerobatics as being a relatively non-essential sport reserved for the lunatic fringe. For that reason, it's quite possible for a Pitts Special owner to find himself on an airport where he is not only the sole sport airplane in existence, but gets a certain amount of heat from the rest of the airport population as being a non-essential member. In some cases, you feel as if they think you are a Hell's Angel (nice metaphor!) sans tattoos.

Exact opposite situations exist at some airports, usually smaller ones in which sport airplanes are the primary population and Pitts pilots are looked up to as being practitioners of an exotic art reserved only for the select few. Obviously, it is better to find the latter than the former.

For the most part, smaller airports are more accommodating to Pitts Specials than larger ones. Not only are they normally less expensive, but the small airport is more likely to have a grass runway as well as a paved one and a Pitts owner could ask for nothing better. Pitts are so much happier on grass, they seem to smile even as they are being pushed out of the hangar. The airplane handles much easier and better on grass and grass-field landings are much easier on tires and brakes. If the grass is smooth, and it has to be to work with the

tiny tires on all Pitts, each landing can become an organic, emotional experience in trying to mate the tires with the very tops of the grass so the airplane just whispers into it.

Don't underestimate the importance of finding a Pitts-friendly airport, since the wrong decisions could put you in the position of having an airplane that has an unhappy home or no place it can go play.

Weather as a Consideration

Another consideration when buying a Pitts, although a minor one, is the local atmospheric and topographical environment; for instance, do you live in an area where it's high and cold or one where it's flat and warm? The stubby little wings that make the Pitts Special the nimble performer it is also make it extremely sensitive to density altitude, something attested to by Pitts owners who fly in the high, thin air of Colorado or similar locations. The same 85 mph approach on a hot day in Boulder is an entirely different 85 mph than seen at Saint Augustine, Florida.

It should also be pointed out that northern latitudes can limit the Pitts pilot's flying year unless he is prepared to bundle up each time he goes flying. Even with the canopy, the Pitts Specials are not known for their passenger comforts, since they have no heating system other than the engine itself. With a canopy, the pilot will feel some heat coming back off the engine but not enough to offset more than about forty-five minutes spent in cold temperatures. However, that doesn't mean you shouldn't fly the airplane during the winter since the airplane performs so much better in cold temperatures. The Pitts pilot just has to get a stiff upper lip (literally) and bundle up for the occasion.

Fortunately, a sister group of the lunatic fringe, the snowmobile and motorcycle crowd, have come up with great clothing in the form of snowmobile suits that make flying the airplane totally bearable in almost any temperature, and it is absolutely worth making the effort.

Why Do You Want a Pitts?

The question of why a pilot wants an airplane such as a Pitts Special is something difficult to answer specifically, but it's a question that should be

answered, if only to determine which Pitts Special the pilot should buy. The first question, of course, should be whether or not competition is an end goal and if so, how serious a competitor he or she wants to be. The answers to those questions will, to a certain extent, determine whether one should consider a single-place airplane for pure overall performance, or whether one should compromise for pure Sunday morning personal amusement and limited competition and buy a two-place.

One of the biggest questions of all begins with whether the pilot is interested in sharing the view or not. The ability to take someone else along either just for the sheer fun of it or for instructional purposes is important. The two-place birds can always be flown as single-place planes, but the single-hole Pitts is about as single-hole as an airplane ever gets, since there is barely room for the person doing the driving.

The two-place airplanes have a lot going for them since even as a single-place they provide a lot of extra room for baggage should they be used for cross-country work. Also, if cross-country is part of the scheme, how often do people actually go cross-country by themselves? In other words, the two-place airplanes offer more utility and are something more than a one-trick pony. Also, if the plane is to be brought into a family situation, it's often a little easier to rationalize it in front of the spouse if it's a two-place airplane, since a single-place looks just a tad self-indulgent.

If serious competition is going to be part of the equation, then the decision has already been made to spend a fair amount of money and either buy a hot-dog single-place or a top-drawer S–2B or S–2S. If the goal is to go all the way into advanced and unlimited, then the decision is made automatically: to play with the big boys takes a big boy machine and that's an S–1S or one of the competition derivitives. Serious competition also means spending a little more serious money than simple sport flying would. However, it still means spending less

A single-hole Pitts develops a lived-in look when it is used in competition, but this degree of wear will not significantly affect the price.

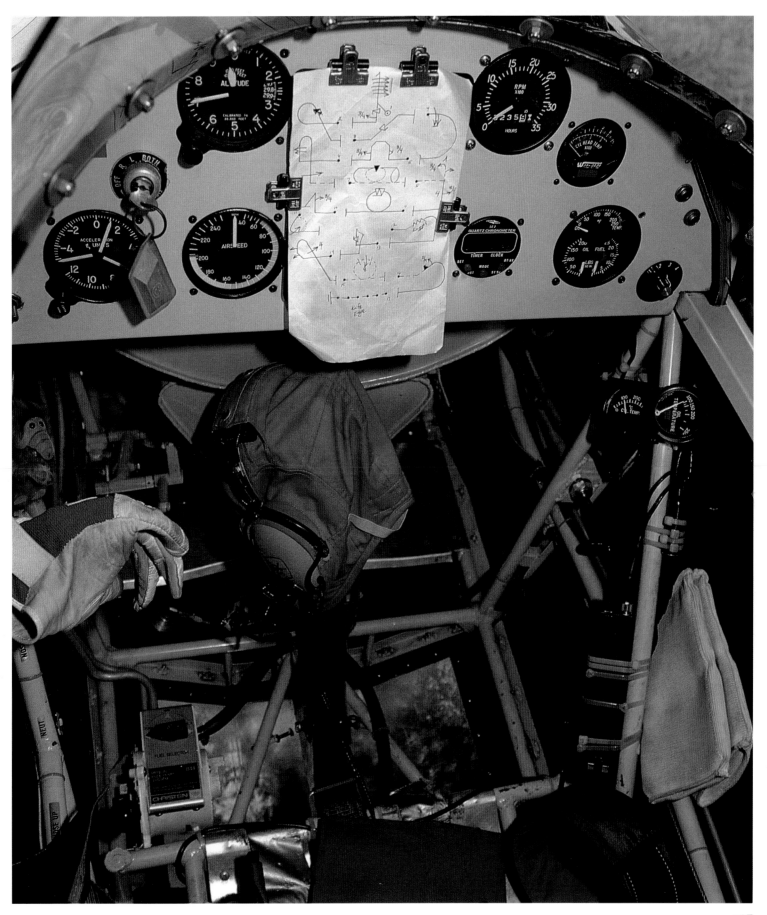

money than for most of the two-place airplanes.

If going out to pull a little g on a Sunday morning is the goal, then there are lots of ways to get into it that don't cost a fortune. A nice little 125 hp flatwing will work just fine, and a 150 hp flatwing would even let the aspiring akrobat work into intermediate category competition. And none of these airplanes would do even half the damage to the checkbook that any of the rest would.

Evaluating Potential Aircraft

Since all Pitts Specials are rag and tube with wooden wings, there are some points of evaluation that are obviously common to all, homebuilt or otherwise. However, the homebuilt is a special case that will be addressed later.

The common points to be examined on all aircraft start with the logbooks and looking for possible damage, most of which would probably be incurred during a ground loop. Ground loop damage would probably be minor, although that's not always the case. Engine time is another major factor which, along with fabric and finish, is as much a matter of flying time and age as anything else.

In the case of fabric, the type of fabric has a lot to do with how long it can be expected to last, with cotton being the shortest-lived and synthetics such as Ceconite or Dacron lasting the longest. Don't even consider buying a Pitts that has a lifetime covering on it such as Razorback, since aerobatic airplanes should be taken apart from time to time to be examined anyway. There has been an ongoing controversy for years as to the suitability of synthetic fabrics for aerobatic aircraft because of the problem sometimes encountered in making the finish adhere to the surface. The finish, whether it's urethane or dope, soaks right into a cotton fabric adhering tenaciously, while the same finish put on Ceconite depends entirely upon a mechanical bond since it can't be absorbed by the fibers.

The finish itself may be split into two different categories, one being the tried and true butyrate dope and the second being any of the newer urethane systems. When evaluating an aircraft, it's important to examine the finish carefully since aerobatics tends to impose loads upon the finish, causing it to crack or deteriorate. This is especially true of

some of the urethanes that are a little stiffer. Look for cracks along the stringers and in the corners of the rib-to-trailing edge intersections. Also examine the belly to see if rocks or other foreign objects have damaged the belly enough to where the finish is beginning to crack or peel. A minor amount of this type of deterioration can be tolerated; however, eventually the finish is going to decide to depart the airplane in flakes or sheets, which will lead to recovering the machine—not a cheap process unless it is done in the cramped sanctity of your own workshop.

The Homebuilt Pitts Special

Officially, the term homebuilt is a misnomer since the FAA actually categorizes them as being in the amateur-built subcategory of the Experimental aircraft licensing category. The Experimental Aircraft Association calls them custom-builts, however, the bottom line is they are all built at home so generically they are called homebuilts. As such, the term certainly is appropos since it clearly indicates every airplane in that category is going to be a little different from the next. There are no detailed rules of the road to guide every builder to do everything exactly the same. Additionally, even if there were such guidelines, every airplane would still be different since each individual wants to build the machine to reflect his or her own individual taste. So each airplane is as different as its builder's imagination.

Each homebuilt Pitts Special must be evaluated on its own strengths and weaknesses. There are, however, some basic areas that should be checked carefully to determine if the aircraft has long-term airworthiness or not:

Welding

The strength of the fuselage is based almost entirely upon the strength of the welds used to hold it together. Grab a flashlight and poke your head down inside the fuselage to look at as many of the clusters down around the wing fittings as possible. The welding should be clean, with the ringlets of each bead somewhat evenly spaced with a more or less smooth appearance. If the welds are wildly irregular, show undercutting at the edges or have a vague roughness showing through to the surface indicating overheating while the welding was

being done, perhaps another airplane should be looked at.

Fittings

As much as possible, try to examine the fittings in some detail since they show the psychology of the builder. For example, if the fittings are clean with nicely rounded edges and display no nicks or scratches, then the builder was obviously detail oriented. If, however, they are a little irregular on the edges, possibly showing some nicks or hacksaw or sanding marks, this is definitely an indication the builder didn't understand what he was trying to accomplish or is not detail oriented. This thought pattern probably carried over into all parts of the airplane, most of which can't be visually inspected without taking it apart.

Materials

Try to determine how many of the materials used, such as wheels and brakes, engine, instruments and so on, were purchased new at the time the airplane was built. If all of the parts came out of an airplane with several thousand hours on it, then for all intents and purposes your airplane has several thousand hours on it that don't show in the logbook.

Documentation, Weight and Balance

Look through the logbooks and the FAA paperwork to find out if the airplane has a permanent airworthiness certificate or one of the newer limited ones. This only means that the airplane will have to be licensed as soon as it changes hands, if it has one of the newer airworthiness certificates. Also look at the weight and balance to make sure the airplane falls within the recommended limits but also—and this is very important—to determine what the airplane weighs empty. Check this against known values of what each one should weigh because a heavy airplane is a poor performing airplane.

In general, a 125-150 hp flatwing should weigh in the 680-690 lb. range. A 180 hp flatwing or roundwing should be approximately 720 lb. These figures don't include an electrical system, upholstery or wet bar. An airplane with lots of upholstery, instrumentation, avionics and so forth is obviously going to be heavy and the performance is going to suffer. The differences in weight determine the entire personality of the airplane.

As a final note in examining homebuilts, try to lay your hands on a set of plans that match the airplane being examined. Use them to make sure the airplane was indeed built as per plans and that there are no obvious deviations that, in effect, make it an airplane other than a Pitts Special. Cosmetic changes such as turtle decks or cowlings are acceptable and to be expected; however, no one should be playing with any of the structural details without knowing what they are doing, and this includes hinging of the ailerons, landing gear design and so on.

Modifications

There are some recognized and proven modifications to the airplane that are generally acceptable, but they almost never add enough to the plane to make them worthwhile. Some of the changes include:

Spar-craft/Stewart Wings

The Spar-Craft wings appeared about the time Curtis Pitts began offering S-1S, roundwings in a completed form (the late 1960s). Spar-Craft wings are essentially bootlegged Pitts roundwings in kit form. They reportedly don't exactly match the Pitts airfoils, the combination of which is patented, and the ribs are cap-stripped plywood rather than being built up. Many are flying, although they are slightly heavier than Pitts wings and don't quite measure up in performance. These are most often seen on the shorter S-1C fuselage, although they also appear on the longer S-1Ss.

Ultimate Wings

Designed by Gorden Price, a well-known competitor and builder, the Ultimate series of wings is an attempt to improve on the Pitts' roll performance, which is already astounding. The Ultimate Pitts does roll faster but doesn't appear to gain anything more in the vertical line. They are also slightly heavier, and many flyers don't like the looks of the squared-off wing tips.

Spring Gear

There are a number of versions of spring landing gear, most of which show up on competition-oriented S-1Ss, although a number of flatwings have also been so modified. This gear is slightly easier to handle on the runway because it isn't nearly as rigid as the original V gear. Its real advantage lies in the lower drag and higher speeds it gives. The gear requires some structural modification to the fuselage, and should be inspected carefully for professional workmanship. Since the Pitts' gear geometry determines its character on the runway, it is critical the gear be properly designed and aligned.

Haigh Tailwheel

The Haigh tailwheel replaces the leaf spring, locking-swivel tailwheel that is common to most light tailwheel airplanes. The Haigh spring is a tapered rod and the wheel is smaller and lighter weight. The biggest difference, however, is that it is a locking unit, meaning a lever in the cockpit locks it solid in a straight-ahead configuration for landing and takeoff. Then, for taxiing it is unlocked and is full swivel. This feature is the subject of some controversy because, when it is unlocked, the airplane is totally dependent on the brakes for steering and can get really squirrely.

Summary

The most important aspect of buying a Pitts is getting out there and looking at a bunch of them. Also, find as many Pitts pilots as you can. Talk to them, get their input, even if it has to be done on the phone. Combine all their experiences and input to form an average opinion of what comprises a good airplane. Also, in the process of networking with other Pitts warriors, it is probable you'll stumble across a number of airplanes for sale that would otherwise be missed.

During the early part of this education program, leave your checkbook at home. Wait until you have the knowledge before spending the money.

Building a Pitts
A Fine Form of Lunacy

Okay, so the Pitts Special bug has bit hard and visions of sugarplum fairies doing outside loops in bright red biplanes fill your head. Somewhere in the cold pre-dawn hours, you bolt straight up in bed and exclaim to your still sleeping spouse, dog, cat or canary, "That's it! I'll build a Pitts Special!"

The urge to build an airplane often strikes otherwise normal folks right between the eyes with unexpected force. Actually, most of the time it strikes them right in the heart—the most illogical, and therefore, most dangerous organ in the body. Then, before they've given it enough thought they find themselves up to their armpits in little pieces of wood, steel tubing and bolts of cloth. It's the guy who gets the urge and then instantly does something about it that falls into the FAA's statistical trap: The Feds' records show only one out of ten traditional homebuilts are finished. This is inevitably because not enough thought was given to the project before lighting up the torch.

When you build your own Pitts you can paint it any color you want. This one displays one of the more imaginative schemes ever to grace a Pitts.

S–1S in the final stages of construction. All the bare portions will be covered with aluminum. The balance is fabric. EAA

To avoid becoming part of the "uncompleted" statistics, several things should be done the second you find yourself walking small circles in your bedroom muttering, "Wanna build, gotta build, gonna build." The first is take an icy cold shower and try to get your nerve ends back to normal. Second, hide your tools. And third, give your checkbook to your spouse, dog, cat or canary. If there is ever a time to get icy cold and logical, this is it.

The process of building an airplane—any airplane—is invariably underestimated by the individual getting into it for the first time. That's unfortunate because building an airplane is long, drawn out and absolutely chock full of opportunities to become discouraged, disgusted and downright frustrated. For those reasons, it is absolutely necessary some serious planning and (most of all) evaluation be done before jumping into the project.

Pre-building evaluations are undoubtedly the most boring part of any project. But evaluation is so necessary in the case of building a Pitts that it should almost be formalized and broken into three basic areas, only two of which have anything to do with the airplane.

The first and most important area to be evaluated is you and the lifestyle you lead. You have to crawl around inside the nooks and crannies of your soul and personality to find out if you are indeed suited to be the builder type. Second, the

facilities and other provisions that you have at your disposal to build the airplane have to be evaluated. And third, and least important, the Pitts itself should be evaluated as a project to see if it fits in with the first two categories.

Self-Evaluation

Examining Your Objectives

In evaluating yourself there are a couple of basic questions that have to be answered, the first of which is: "Why do you want to build a Pitts in the first place?" If, as you stand in front of the mirror looking yourself right in the baby blues and ask the question, the answer comes back "because I want to fly a Pitts," then you have just hit upon the worst reason there is to build a Pitts.

The problem with wanting to *build* an airplane simply because you want to *fly* that airplane is you tend to overlook, and therefore don't truly understand, the thousands of hours that will be spent in a dark cubbyhole. A huge amount of time will be spent creating this brilliant flying machine and if before you start, you have already mentally strapped yourself into it and gone trundling off over the horizon, you'll be in trouble from the beginning. If flying is your prime motivation for building, there will be a tendency late at night, when you're nose down in a pile of tiny widgets that don't even remotely resemble an airplane, to lift your head and

The framed up skeleton of a single-hole, S-1C shows the relative simplicity of the design. EAA

daydream. You'll look far off in the distance at the time when you will be flying your airplane. The more you do that, the more discouraged you will become because the further you get into the project, the more obvious it will become that your cherished dream of cavorting in a warm sunset is years down the road. Many dreams have died simply because the long stretch of time between the dream and the reality makes the warmth of the idea go away. When the first excitement diminishes, the urge to build atrophies until a classified ad offers the project for sale.

If you look yourself in the mirror and ask, "Why do you want to build," and your answer has to do with the money to be saved, then you ought to sit down with a pad of paper, a pencil and a $4 calculator.

It is very enlightening to add up the total cost of building an airplane versus buying one that's already flying, or partially built. As a rule, almost any variation of Pitts is available for little more than the cost of the materials. Often the less-expensive flatwing varieties with the smaller engines can be bought for not much more than the retail cost of

the engine and the instruments or avionics.

The only financial benefit to actually building the airplane yourself is you can precisely regulate the rate at which the money goes into it. The Pitts Special is one of the more easily controlled cash-flow projects because the materials can be bought separately and in increments that exactly match your pocketbook. For instance, the initial $400 spent on tubing will let you build the fuselage and probably most of the tail and keep you busy for quite a number of months before you have to spend any additional money. You can even get into it slower than that and drop $75 to first build the tail and so forth. Same thing goes for the wing. Yes, the wing materials may run as much as $1,200, but who says you have to build all the wings at one time, and who says you have to buy all the materials at one time? A hundred dollars worth of ¼ in. cap-strip material and ¹⁄₁₆ in. plywood will put you into the rib-making game and keep you off the streets for a month or two until you have to spend money for the spars and drag or antidrag wires. Even then, you can buy just enough material to build one wing panel at a time.

Financially, the thing that can most accurately be said about building a Pitts is you won't save any money in the long run, but in the short run it may enable

you to get into the Pitts Special game without dropping a big bundle of cash up front.

At this point someone's going to ask, "Are there any really good reasons to be building?" And the answer is "absolutely!" The primary and most important reason to be building a Pitts Special (or any airplane for that matter) is that you simply like to create. Sculpting formless material into something that not only flies, but has a life of its own, is a tremendously satisfying process which is at least as powerful as actually flying the airplane. Just think of it: Every piece of the material required to build a Pitts Special, excluding the engine, instruments and wheels, could be easily fit into a box 1 ft. square and 12 ft. long and still have space left over. There is no more creative process than that of pulling that box into your workshop to have it emerge several years later as a shining example of not only something that flies, but something which mirrors your own creative personality and ability to face challenges.

Building airplanes is a unique endeavor. There are few things in the world an individual can craft with his own hands that has as much unexplainable emotion attached to it. An airplane is a living, breathing thing that can take its builder-pilot to places undreamed of and change his life forever in ways the average man could never understand. In building an airplane, the builder has, in effect, written his own script for a new chapter in his life that is suited only to himself. This is because only he knows exactly how he wants this airplane to fit him and how he wants it to look. Only he knows how best to satisfy that indefinable yearning for something that is uniquely his.

One other more mundane and less ethereal reason to build an airplane yourself, rather than buying one, is that crafting every piece yourself is the one and only infallible way of knowing exactly what went into the airplane. As mentioned in the last chapter, homebuilt aircraft vary wildly from one builder to the next, so the quality of the airplane also varies. At least in building it yourself, that unknown no longer exists.

Evaluating Yourself as a Pilot

I've said it before and I'll say it again, anyone with any background and any skill level can learn to fly a Pitts Special *if the individual really wants to.* In this case, wants to automatically says that individual is willing to rise to the challenge presented by the airplane's handling characteristics, and approach it in such a way he or she will enjoy that challenge. This does not necessarily describe all pilots.

Many pilots like the idea of flying what is essentially a "hot airplane." However, when it comes down to reality, they may find they don't enjoy the apprehension that goes along with working at the outer edge of one's talent envelope while involved in the learning process.

The question of where a person fits on the pilot totem pole is something that has to be looked at carefully: If you like flying to be a no-sweat, no-demand activity, then you're barking up the wrong airplane tree with the Pitts Special. If, however, you enjoy making each flight an experiment to improve yourself as a pilot, then the Pitts Special is the finest aeronautical tool you'll ever lay your hands on since, regardless of the type and flavor of Pitts Special you build, it will always have something in reserve to teach you and you will never stop learning.

Since we are talking about building a Pitts, the pilot-builder is almost automatically talking about a single-place airplane. This in turn means the checkout procedure is even more of a challenge, since he must face the ultimate test alone. Fortunately, with so many two-place Pitts available, he can, at the very least, sit in the front seat and approximate what he will see in a single-place plane. Thanks to the two-holers, checking out in a single-place Pitts is no longer quite the adventure it once was.

Evaluating Your Lifestyle

As it happens the Pitts Special, once it's finished, will fit into almost any lifestyle because it's quite a low-demand airplane on your time. It takes only a few minutes to do some serious yanking and banking, and chances are you aren't going to go running off to Bermuda or the Bahamas in it. The operative phrase here, however, is "once it's finished." Before it's finished, the Pitts Special, like any airplane-building project, is incredibly time-intensive. For it's size, the Pitts is much more time-intensive than many of the new composite kit planes available.

In side view the compactness of the design is in evidence. S–1S pictured is in the EAA shop, hence the Stuka tail in the background. EAA

Assuming the individual builds the entire machine himself, he can count on having approximately 3,000 hours tied up in the project from start to finish. Think about that—3,000 hours. That is three hours a day, every day of the week, for three years. Do you have that kind of time? Better yet, is your family structured in such a way they'll allow you that kind of time? Time is almost never free. It comes from someplace else. It either means television isn't being watched, tennis matches aren't being played, sleep is being avoided or, as is often the case, at least part of the time comes from that which would normally be spent with the family. Is that going to be a problem? More than one airplane project has been abandoned because the amount of time it required was at the expense of the family relationship, and the builder wisely opted to give up the airplane rather than his family.

The availability of time can't be underestimated in importance. On the other hand, if you can only get five hours a week in, that's perfectly OK. More than one Pitts has taken fifteen years to build.

Determining Your Skills as a Builder

There is often the mistaken notion that airplanes are extraordinarily complex, precise structures requiring superhuman understanding and arcane skills, some of which include a little alchemy and black magic. That is not the case! Building a Pitts Special, as with building most other airplanes, uses basic skills that have been around for at least three quarters of a century and can be mastered by anyone who can

sheetrock a wall or build an accurate set of bookshelves into the closet.

Just as anyone can learn to fly the Pitts Special, certainly anyone who doesn't already have all, or part, of the required skills can easily learn them. The primary skills needed are welding and woodworking, with secondary skills being those required to apply the fabric and finish.

Most often, the most serious skill that has to be mastered is a mental skill known as project orientation. Without it, a builder's hands may be the most talented in the world, but the project, in this case, a Pitts Special, will never be finished. Project orientation is a mindset in which the project is kept in a particular cubbyhole of that person's mental makeup so it is always there and always being worked on. Many successful builders say their approach to the project is making sure they physically touch it every single day—even if it only means sweeping up the shop or tightening a single bolt. They want to keep it in their mental mainstream and therefore keep hammering away at it.

A major part of project orientation is the feeling of tangible progress that comes from completing a part. Often that means organizing the project in such a way that when only fifteen minutes are available, fifteen minutes' worth of solid progress can be made, either in making a tiny project, or putting the finishing touches on one that was done yesterday during that day's fifteen-minute period.

Another personal trait that must be evaluated or developed is the person's attention to detail, since that is the characteristic which separates craftsmen from those who would be better off puttering in the garden. It is essential the builder understand and remember constantly what it is he is building: This is a machine that gets him high enough to bust his butt. It is a machine that is designed to take him into an environment for which he is personally, totally ill-suited and in which he absolutely cannot survive without the machine—he's not a bird, and is dependent on the machine for his wings. Therefore, good enough is simply not good enough. There cannot be any feeling that corners can be cut, because cut corners can cost your life. Yes, there are some parts of the project that are more important than others, but the mental attitude has to be such that all pieces, all parts, all phases of it are important.

It's the unending attention to detail that means the edges of all the fittings will be clean and straight. There will be no scratches tolerated. All the welds will be even and as carefully formed as possible. Alignments will be measured in thousanths of an inch rather than sixteenths. In almost all cases, that amount of detail is not necessary. However, not being the designer, the builder won't necessarily know where his attention to detail will pay off. Nor will he know where the absence of attention to detail can cause a potentially dangerous situation. Therefore, an unending effort to make every tiny part perfect will result in a totally safe, absolutely perfect airplane.

As part of the mental project orientation, it is both important and helpful if each little piece being worked on at any given time is looked at as a project unto itself. For the time being, that part must be disassociated from the airplane, which means the airplane becomes nothing more than a long series of little projects, rather than one big one. By maintaining this mindset, the builder is far less likely to become discouraged because he will be seeing solid progress as he finishes each little project and puts it into the pile marked "finished."

When all the little projects are attached to one another and are laying in the finished pile and there are none in the to-be-done pile, the airplane will be finished.

Facilities and Other Provisions

Far too often, builders think of workshops in terms of space, rather than other, possibly more important factors. A workshop that helps the project, rather than being an obstacle in itself, doesn't just happen, it is designed with a goal in mind and, as such, is a project in itself. In the first place, when setting up the shop it is critical every effort be made to overcome the psychological obstacles that kill projects such as Pitts Specials. The largest one of those is the cast iron that tends to develop in everyone's butt shortly after dinner. The first half hour after dinner is the critical decisive factor in any project. If the individual is able to regularly force himself into the shop during that period, the project will eventually be finished. If he's not and he sits down in front of the television set to wake up an hour and a half later, when the dog is scratching because he wants out, then the project is doomed. Therefore, the workshop must be such that it practically sucks the builder into it after dinner because it has attractive features other than the airplane itself.

The primary features of a workshop that act as magnets to consistently draw the builder into the shop are combinations of space, proximity and a user-friendly atmosphere.

Space, or the lack thereof, frustrates a lot of would-be Pitts builders. However, there is space, and then there is space and the problems of finding or developing it vary wildly. For instance, a dweller in midtown Manhattan has an entirely different problem than his compadre out in the suburbs with a two-car garage and two acres of backyard. Yes, there have been many airplanes built in back bedrooms and living rooms, but that is, by far, the exception rather than the rule. It's a simple fact of airplane life that the availability of space has as much to do with the spawning of airplane projects as anything else. Airplane-sized spaces should have airplanes in them.

Fortunately, Pitts Specials are the airplanes to be building if space is at a premium. Compared to almost any other normal airplane, Pitts Specials and their various body parts are so small, they fit into just about any workspace. However, even in the Pitts' case, there *is* such a thing as construction claustrophobia. Just because the airplane fits into the hole doesn't mean it's the right kind of hole.

The question of how much space is enough often comes up, and it could be answered that there is no such thing as too much space. However, since all this space has to be lighted and heated, it could also be said yes, there is such a thing as too much space. That, to put it mildly, is usually not the case.

Since the largest single part on a Pitts Special is the top wing (17 ft. long by roughly 4 ft. wide), it's obvious that part can easily be worked on in a single-car garage. However, a single-car garage should be considered the rock-bottom minimum, since there's almost no room left over for work benches or storage space. The ideal size workshop is a two-car garage, which is roughly 22x22 ft. That space gives plenty of room to store the parts once they're finished, as well as having plenty of room to construct benches and to properly arrange the tools. If each part is laid out by itself—for example, if the top wing is laid beside the fuselage—there should be a minimum of 3 ft. around each of the parts so as to allow easy access. If you're constantly having to step over and around things to get to the other side of the part in question, simply navigating around the shop becomes a hassle.

It should be pointed out a workshop should be a workshop—not a cleared space in the middle of the residue that seems to collect around a household. A workshop should have no place for bicycles, rakes, lawn mowers, snow blowers or any of the other household support items that grow with the passing years. A workshop is a workshop is a workshop, and everyone in the family should understand and respect that.

A major portion of your life for the next three years is going to be spent in that workshop, so it only makes sense to put off actually working on the Pitts until the workshop environment itself is everything it should be. Certain items are critical. These include lots of light, well-finished walls and ceiling, storage for parts and the correct number and size of work benches.

Light is certainly one of the most important environmental aspects, not only in being able to do the work, but also in keeping the shop bright and cheery. It is also one of the easiest things to correct because used fluorescent light fixtures are readily available. By watching the local used merchandise newspapers and haunting electrical contractors, it's quite easy to come up with multi-tube 4 and 8 ft. fluorescent fixtures that have been removed during the course of a renovation. They're cheap and most of the time have fluorescent tubes in them. The workshop ceiling should be practically covered with fluorescent tubes. In addition, take a 4 ft. unit, make a stand for it and put an extension cord on it so it can be plugged in and moved to wherever additional light is needed.

If the workshop has unfinished walls, some serious thought should be given to both insulating those walls and sheetrocking them. Sheetrock and the associated furring strips or two-by-fours are not at all expensive and they contribute mightily to the usability of the workspace. Among other things, sheetrock gives a fire barrier in case the workshop is part of the house. It also creates a wall surface that is reflective and adds significantly to the overall cheeriness of the work area. There is one other aspect to spending some time sheetrocking your shop—it builds good work habits. If you can sheetrock a wall so it is perfectly straight with no waves, then you've had your first lesson in building accuracy.

Parts storage should be arranged as far out of the way as is possible. This should include racks around the tops of the walls and under any benches that are built. You also have to remember—as the airplane is being built, you will come up with more and more big pieces. Wings may start off as a small pile of tiny parts but they rapidly grow into large components that have to be stored and protected.

Work benches are the heart of any airplane because they form the datum surface that will determine how true and accurate the structures will be. You'll need a bench that is long enough and wide enough on which to construct a complete fuselage, and then you'll need one that's a different dimension to

use in jigging up the top wing. The top wing bench can actually be two 4x8 ft. benches bolted together lengthwise, while the fuselage can be built on two 2x8 ft. benches. Therefore, one of the 4x8 ft. benches should be constructed in the following manner:

Make up two frameworks that measure 2x4 ft. and bolt them together side by side down the long dimension. Then skin them on the top with a single piece of 4x8 ft. ¾ in. thick plywood. By cutting the surface lengthwise at the 2 ft. dimension, you'll have two identical benches and the surfaces will line up because they were skinned at the same time. Therefore, those benches can be put end to end to make the fuselage jig bench, and put side by side to become part of the wing bench. These benches have to be positively straight and level and should be stout enough to resist any tendency to move around either under weight or humidity changes.

Not all airplanes are built in the sunny south, nor are all of them built in warm, comfy workshops that are part of the house. For that reason, heat is almost always a major consideration for homebuilders. If it's not possible to tie into the house heating system, the varying approaches include spending the money on a first-class heating system, although a large percentage of builders prefer to go a more economic route. Certainly one way is to haunt the local heating contractors and see about picking up a good used oil-fired hot-air furnace from a renovation. In most cases, these are hooked up with nothing more than an on-off switch to blow heat into the workshop when it's needed.

Another approach is the ever-present kerosene-fired Salamander, as seen in every construction site north of the Mason-Dixon line. These are fine, but you must remember any kerosene-fired heater has an open flame which greatly reduces the ability to use it when working with flammable paints or finishes of any kind.

One interesting approach to putting additional heat where it is needed is to pick up a couple of 4 or 8 ft. electric baseboard heaters and put a 220 volt, surface mount plug on an extension cord so they can be plugged in and placed anywhere you're working.

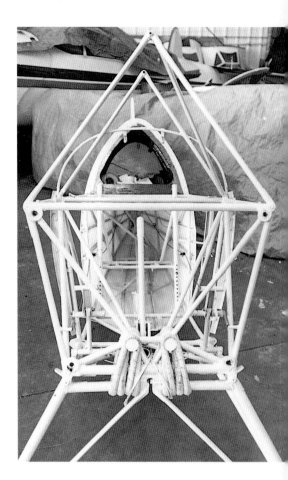

The S–1 series fuselage is trapezoidal to give maximum shoulder width in the smallest area. Note the bungees on the landing gear.

Tools—How Many Is Enough?

The Pitts Special is a simple airplane. No, better than that, it's simpler than simple. It borders on being erector set or blacksmith simple, which means, with the exception of a welding torch, the airplane could be built entirely with normal hand tools. However, that's doing it the hard way. Because of the amount of steel to be worked and the low-cost nature of many power tools, it only makes sense to make an investment in some basic power tools which will make your life a thousand times easier and, therefore, further guarantee the chances of the airplane being finished.

Probably the best place to locate tools is to buy them used through the local want-ad press type of neighborhood resale magazines or newspapers. Everything that is needed can be found there and at bargain prices. However, all of the following tools can be bought brand new

125

in the form of Taiwan imports, and they are perfectly capable of doing the job for a total cost of under $500.

Ranked in order of importance, certainly an 8 in. bench grinder would be number one since that would allow you to dress the ends of tubes as well as doing the roughshaping of fittings and other pieces of metal. Right behind the bench grinder in importance would have to be a drill press which will allow you to drill consistent and accurate holes both in wood and metal. Next in line would be the 6x48 in. stationary belt sanders which can be used horizontally and vertically. If one of the Taiwanese metal cutting bandsaws is also purchased, the builder will find that between the bandsaw and the stationary sander, it is almost as if he has his own machine shop. He will also find that his ability to produce perfectly shaped fittings will not only be greatly enhanced, but the time required will be reduced by 90 percent. The builder only has to spend about twenty minutes with a hand hacksaw cutting a fitting out of 0.063 in. thick chrome-moly steel to realize how nice one of the $200 bandsaws is. By spending $500 for the aforementioned tools, the builder will not only move much faster on the airplane, but will also find the tools usable in every facet of his life. In other words, the right power tools are a good investment.

The only true specialty tool the Pitts builder cannot do without is the old-fashioned oxy-acetylene welding torch. When I say old fashioned, don't assume I'm talking about the big torch handle and blow pipes you've seen someone building a bridge with. The only welding torch that's truly applicable to aircraft building is the Smith Industries Airline series of welding torches. This is a small, extremely lightweight torch which makes it much easier to get into the small areas an airplane is crammed with. The torches, along with a set of regulators and tanks (which will probably have to be rented), combined with basic tools gives the builder the ability to build his own Pitts Special with little or no help from the outside.

The Pitts as a Project

The Pitts Special, as I have mentioned, is an extraordinarily simple airplane. It has a few characteristics that are unusual but nothing major, and every aspect of the airplane can easily be handled by a backyard builder, with the proper tools and attitude. The following paragraphs give brief descriptions of the various components of the airplane and what would be involved in construction:

Fuselage

The fuselage is steel tube welded up and is vaguely trapezoidal in shape, that is, it's wider at the top than it is at the bottom. This makes the construction of a Pitts fuselage slightly different from most steel-tube fuselages. The normal method for a Pitts is to build it upside down with the top surface flat on the work table, rather than doing the sides first, as with most similar airplanes. This requires some vertical jigging to properly locate the bottom longerons in space, which requires a little more attention to detail to ensure accuracy.

Tail Surfaces

The horizontal and tail surfaces are made up of bent leading edge tubing with a straight tube rear spar. There is no internal structure other than the ribs which are bent up out of 0.025 in. thick 4130 sheet steel. These ribs can be bent up on a bending break or formed around tapered steel form blocks. The ¾ in. thick tubing forming the leading edge of the horizontal stabilizer is normally bent cold around a plywood form.

Wing

The Pitts Special wing is quite literally nothing more than a model airplane structure with a thyroid problem—it is much bigger. The only unusual aspects stem from the fact that the top wing is one piece, which requires an accurate splice in the center. In making that splice, the inboard spar ends must be dressed to the proper sweepback angle so as to accept the splice block. The ribs are all the same size and built up out of ¼ in. square cap-strip material with ¹⁄₁₆ in. gussets at the intersections. The drag or antidrag wires which crisscross throughout the wing are drilled completely through the spars, with nuts on the outside faces of the spars. This arrangement eliminates the need for drag or antidrag fittings, which is much simpler and reduces the cost as well.

Flying Wires

The aircraft uses a number of flying and landing wires both in the wings and on the tail which not only have increased in cost rapidly over the last few years but can have a significant lead time in ordering them. Therefore, both the cost and the time factor have to be planned into the total project.

Engine

In purchasing the engine, both the size and the number of hours on the engine determine the cost which, in turn, drives the cost of the entire project. All airplane engines are expensive, it's just that some are ridiculously so. And there is not that much overall cost difference between a 150 and 180 hp engine, although there are probably more used 150 hp Lycomings for sale than any other engine in aviation. The 150 hp Lycoming is also one of the most dependable, although both it and the 180 Lycoming are considered to be bullet-proof motors. The amount of money spent on the engine is totally up to the whims of the builder, since a new motor will cost anywhere from two to three times the cost of a mid-time used one, and the overhaul cost on any four-cylinder Lycoming from a top-quality shop is always going to be approximately the cost of a medium-level Chevrolet. Like I said, they are all expensive.

Cost of Building a Pitts

The cost of building a Pitts Special today, including the tubing, the wood and all materials to construct the complete airframe minus the engine, would be approximately $8,000 to $10,000, which includes a reasonable cost cushion. The engine could run anywhere from $6,000 to $20,000, depending on the aforementioned variables. Using the automotive yardstick as a floating inflation standard, we can estimate that the airframe minus the engine can be built for less than the cost of a bottom-line Chevy, while the whole airplane is going to consume parts that range from the absolute top-of-the-line Chevrolet to the bottom-line Mercedes. Using these figures, the cost of the airplane can be calculated any time in the foreseeable future with no regards to inflation.

Building versus Buying

There are a couple of alternatives to building the complete airplane from scratch. One is to contract with any of the numerous builders specializing in Pitts Specials and have them build all or part of the basic airframe for you. In that situation, you are trading your time for a reasonable amount of money. For example, a custom-made fuselage frame, not including the tail, will cost you approximately eight to ten times what the material alone costs. Only you can be the judge as to whether that tradeoff is worth it.

A second possibility is to buy a project that is partially completed. In most cases, this would mean buying an airplane that's already on its gear with a fuselage welded and ready to have the parts attached to it, and the wings probably finished as well. This has the advantage of buying it for approximately 50 to 100 percent of the material cost alone, with someone else's labor thrown in free. This of course requires an intense inspection of the airplane to verify the craftsmanship as being something you would be willing to live with. However, since the airplane is generally uncovered, that type of inspection is easy and quick to perform.

In looking for partial projects, one basic fact should be understood: they are always on the other side of the country from you. If a partial project shows up in the trade magazines and is attractive to you, try to find someone in the area who can make the first inspection to tell you whether or not you're looking at a piece of junk that's best converted into a jungle gym, or something that's really worth your while. Then, assuming you don't have your own trailer, you are going to have to rent one on-site and build jigs to cradle the wings alongside the fuselage for the trip home. This cost and the time involved has to be factored into the purchase cost of the project.

At any given time, there are plenty of projects for sale and it's just a matter of waiting until the right one comes along, and then investing the time to look at it.

Cost versus Return

As I pointed out earlier, the selling price of a Pitts depends greatly upon

where it falls in the competition scheme of things. A small-engine flatwing is going to be worth only about a third of a 180 hp S-1S, assuming both are built to approximately the same quality. It's important this difference be noted, since it costs almost exactly the same to build a flatwing from scratch as it does to built a roundwing. The financial return, however, varies greatly. In addition, it can be assumed an S-1S which sports all of the competition modifications such as spring gear and symmetrical ailerons is going to command top dollar. Unfortunately, the finest-built 115 hp Pitts in existence is going to be a very difficult airplane to sell, if only because it doesn't live up to the fire-breathing image Pitts Specials engender.

An S-1S takes up practically no room in a shop. A 180 Lycoming is mounted on this one, which is typical.

When building an airplane, it must be understood that when it is resold only the top-dollar airplanes, such as the S-1S competition airplanes, are going to return anything more than the parts cost. In those cases, they may pay you back as much as $4 or $5 per hour for your time invested. The lower cost airplanes will resell at almost exactly what their parts cost, and your labor will be thrown in for nothing. In situations like that, only you, the builder, can make an evaluation as to whether building an airplane's a good idea or not.

Index